# REACTOR ACCIDENTS

*Nuclear Safety and the Role of*
*Institutional Failure*

David Mosey

Nuclear Engineering International Special Publications
in conjunction with
Butterworth Scientific Ltd

First published 1990

© Nuclear Engineering International Special Publications, 1990

**British Library Cataloguing in Publication Data**

Mosey, David
    Reactor accidents: nuclear safety and the role of institutional
    failure. - (Nuclear Engineering International Special
    Publications).
    1. Nuclear power industries. Accidents
    I. Title II. Series
    363.1799

ISBN 0-408-06198-7

**Library of Congress Cataloguing in Publication Data**

Mosey, David
    Reactor accidents: nuclear safety and the role of institutional
    failure / David Mosey
       p.    cm.
    ISBN 0-408-06198-7 : £18.00
    1. Nuclear reactors — safety measures. 2. Nuclear reactors —
    accidents. 3. Nuclear power plants — management
    I. Title.
TK9152.M68   1990
363.17'99—dc20                   90-1486
                                     CIP

Cover design by Sarah Mercer
Electronic typesetting by BeesNees, Chessington, Surrey.
Printed and bound by Westow Press Ltd, London SE19.

Nuclear Engineering International Special Publications, Quadrant House, Sutton, Surrey SM2 5AS, UK

# ACKNOWLEDGEMENTS

That this review of seven nuclear accidents has actually appeared is very much due to the determination and forebearance of NEI's Andrew Cruickshank whose patience with, and encouragement of, a procrastinating author has been very much appreciated. Other significant help from NEI came in the form of James Varley who contributed the review of the Chernobyl accident. Toronto nuclear safety consultant Keith Weaver has allowed me to plunder shamelessly his ongoing work on institutional failure and provided much valuable comment, particularly on the NRX and SL-1 sections. A very large number of friends and colleagues at Ontario Hydro (particularly in the Nuclear Safety Department) have also provided help, encouragement and much spirited discussion. However, any errors or omissions in this work are my own, as are the opinions and interpretations expressed.

David Mosey
Toronto, September 1989

# The Author

David Mosey was born and educated in the United Kingdom. In 1970 he went to Canada and joined the Technical Information Branch of the Chalk River Laboratories of Atomic Energy of Canada Ltd. In 1974 he moved to the Canadian National Research Council and became involved in its energy programme. In 1976 he moved to the electricity utility Ontario Hydro as a technical writer, and in 1979 became nuclear information officer. In 1984 he joined the Nuclear Safety Department of Ontario Hydro where his principle area of interest has been the study of high consequence accidents.

# Reactor Accidents

# Contents

# INTRODUCTION

It is said that the Roman Emperors maintained a particular functionary to hide behind the throne and whisper into the Imperial Ear: "Remember, Caesar, thou art mortal". It is also recorded that Bernard Law Montgomery had hung on the wall of his caravan a photograph of Rommel. And in Canada (at least) it is reputed that graduating engineers are presented with a ring made from the metal of a structure that has failed. However apocryphal, these examples illustrate the principle, which at least receives widespread lip-service, of the recognition of human fallibility and the benefits of learning from the past — especially past mistakes.

Review of past nuclear accidents is not simply an academic exercise for the historian of science, but should be central to developing our understanding of the challenges to the safe and economic operation of nuclear reactors.

In 1979 and 1986 the accidents at Three Mile Island 2 and Chernobyl 4 were subjected to intense technical scrutiny, and both events have generated a large body of literature. Older accidents have, with the passage of time, become less immediately accessible. Yet review of these older events can not only remind us of some of the basic principles of reactor safety but also offer fresh insights. An example of this was the publication, in 1979, of the classic Levenson and Rahn study of the consequences of severe reactor accidents where the authors reviewed old accidents and destructive tests to illustrate some fundamentals about the degree to which natural mechanisms act to limit the movement of radioactive materials from reactor fuel to the environment.

A review of some of some of the classic reactor accidents can thus be more than an exercise in technical nostalgia. These accidents involved considerably simpler machines than those at Three Mile Island or Chernobyl and their examination brings one face to face with some of the fundamental technical dimensions of nuclear safety.

Perhaps more importantly, one can also see very clearly the necessity for an appropriate institutional structure which can support the technical dimension and which can evolve and adapt to changing circumstances. A good example is the NRX accident of 1952 which not only led to the formulation of a fundamental technical principle of nuclear safety in the Canadian reactor development programme — provision of a completely independent, fast acting shutdown system — but also identified the need for the development of a formal operational safety philosophy and the institutional structures to support it.

A modern nuclear power plant is an extremely complex collection of systems. When an accident such as TMI or Chernobyl is examined there exists the danger that we can become diverted down a technical by-way through examination of detailed differences (or similarities) between nuclear power plant systems. The search for fundamental understanding may be distracted by an accurate (but nevertheless peripheral) comparison of such things as boiler water inventories or containment provisions. Such comparisons may indeed be of significance but should not be allowed to obscure the fundamental causes of an accident. If they are allowed to do so there is the danger that a scholarly assessment can degenerate to an "our toy is better than their toy" polemic. At its best, this can impair that goodwill and mutual respect between nuclear professionals in different institutions or countries which is essential if technical information exchange is to continue and flourish. At its worst it can engender a baseless complacency which is a most inappropriate intellectual climate for the healthy technical and institutional evolution of nuclear safety.

Review of old reactor accidents offers much less opportunity for distractions. Another reason for reviewing such events is to clarify the record. Discussion of recent accidents such as Chernobyl quite often invokes brief references to older events and such references can sometimes be incomplete and hence misleading. A minor example of this problem appears in the 1988 Uranium Institute publication *The Safety of Nuclear Power Plants* whose list of accidents includes Fermi-1 and SL-1. The short descriptions of these events do not seem completely to reflect their nature or their proximate causes.

## Seven reactor accidents

What has been attempted in this collection of essays is to examine seven of the most frequently cited reactor accidents, including Chernobyl and Three Mile Island. All seven involve core damage of severity ranging from local fuel melting (Fermi-1) to total destruction. Only two of these accidents resulted in immediate fatalities (SL-1 and Chernobyl) and only two (Windscale and Chernobyl) had any significant offsite radiological consequences. Three accidents involved reactivity transients (NRX, SL-1 and Chernobyl), two involved local coolant flow blockage (Fermi-1 and Lucens) and two involved severely degraded core cooling (Windscale and Three Mile Island). In two cases (Fermi-1 and NRX) the reactors were returned to service.

These seven accidents all seem to be "much talked about" both within the nuclear industry and in the general body of literature related to discussion of nuclear energy issues. Indeed, it was this feature which prompted their selection. Space and time constraints precluded discussion of accidents which, as the appendix to this chapter suggests, in terms of technical interest and significance are certainly comparable with the seven treated here.

It should be emphasised that the extent and significance of releases of radioactive material resulting from these accidents will not be discussed ... the literature on this topic is extensive and exhaustive. What will be explored will be the events leading up to each accident and the accident itself.

It is very important to emphasise strongly that there is no intention in this work to ascribe "blame" or to impute incompetence to any individual or organization involved in any of these accidents. It is also important to resist the temptation, as we look back a quarter of a century or more, to dismiss errors as "obvious". Hindsight is deceptively reliable and, before consigning past accidents to the "obvious" wastebasket, it would be well to remind ourselves that opening the air bypass valves at NRX in 1952 — a key event in that reactor's accident sequence — points to a failure as "obvious" as leaving feedwater isolation valves closed at TMI some 27 years later. It is a salutary exercise to ask ourselves how people in 2020 might view current nuclear power plant performance, which is certainly not error-free.

Internationally the nuclear industry has a record of safety in its engineering and operations which can only be described as outstanding. That this is so must be in part due to the candour and clarity with which accidents and incidents are reported and the rigour with which they are analysed. The original documentation for all the accidents discussed here ranges in quality from excellent to outstanding. The original report by Sir William Penny and others on the Windscale accident (released recently under the UK's 30-year rule), for example, is a model of clarity, candour and elegance. The nuclear industry has learned much from its accidents and it is possible that these past events have still more to teach us.

## Institutional failure

In the course of discussion of these accidents the term "institutional failure" will occasionally appear. This was formally defined in a paper presented by the author and Keith Weaver at the 1988 Canadian Nuclear Society's Annual Conference as "the impairment or absence of a corporate function which is necessary for the safety of an installation. Such a failure is the result of human error in activities which may not be acknowledged as important to safety and may occur far from the man-machine interface". Review of a number of nuclear and non-nuclear accidents seems to reveal this class of failure as a significant causative factor.

Institutional failure has its origins in a failure to recognise the long established principle that operational safety in any technology is not solely the responsibility of those at the man-machine interface — that is, the operating institution has clearly definable material responsibilities for safety which may not always be fully understood or adequately discharged by the management of the institution.

It could be objected that the concept of institutional failure is simply a superfluous artifact and that management actions cannot be evaluated as "correct" or "incorrect" in the same way as, for example, operator actions. In the management of any enterprise, it could be argued, decisions have to take into account a vastly wider range of considerations than are encountered in a control room, and the only criteria against which such decisions can be measured are the economic performance of the corporation, its internal policies and the various laws and regulations to which the activities of the corporation are subject.

Such an argument, however, should be countered by noting two things: first, the success of an organization is absolutely dependent upon the safe operation of its equipment; and second, any assumption that the safety of any technology depends only upon the skill and dedication of those at the man-machine interface is questionable. In fact, the dependence of safety on a range of functions was recognised quite early in the history of industrial development, most notably in the development of the railways in Nineteenth Century England and the evolution of railway safety regulation and legislation.

Indeed, railway development in the nineteenth century offers some interesting parallels with nuclear industry development this century in that:

❏ The technology was new, large scale and had the potential for high consequence accidents.
❏ Unprecedented demands were placed on materials, design and the management of operations.
❏ Early government regulation was felt to be necessary for the protection of public safety.
❏ The technology aroused intense public controversy.

Even in their earliest days the railways offered transport that was cheaper, faster and (on a passenger fatality rate basis) almost three orders of magnitude safer than stage coach travel.

The earliest railway safety legislation in England (the 1840 Regulation of Railways Act) not only established the Railways Inspectorate but also required, for example, that all railway lines provide fencing of the permanent way and control of level (grade) crossings. These provisions were an explicit recognition of the fact that maintaining safe operation of this technology was beyond the capabilities of a single person or group of persons (the train crew). The operating institution had the responsibility to establish and maintain certain functions and equipment. The 1889 Regulation of Railways Act included the most extensive list of specific requirements, including the specification of the braking systems to be used on passenger rolling stock, the operating principles to be used and the requirement for interlocked points and signals. The Railway Inspectorate's authority was also widened and this body issued a number of its own standards, including quite specific requirements for passenger station layout and design and marshalling yard configuration.

As Rolt documents in his classic history of the evolution of British railway safety, *Red for Danger*, there were numerous instances when the Railway Inspectorate, who had the responsibility to investigate and report upon railway accidents, felt that the operating institutions failed to discharge completely or successfully their safety responsibilities. These failures were sometimes identified quite bluntly. For example, a collision in 1870, which resulted in five deaths and 57 injuries, was immediately attributable to the fact that points had been left set incorrectly. The Inspector stated in his report: "I find the company's management wholly to blame for this accident", supporting this by noting that the railway company's attention had been drawn some seven years earlier to the specific technical deficiency which allowed such incorrect setting, that technical means for obviating this hazard had been available for 14 years and that such means had been required on new railway lines for ten years.

It is important to note that while the proximate cause of the fault condition (incorrectly set points) could be called "human error" the underlying cause was the failure of the institution to provide proper equipment and systems.

Another particularly interesting example was the Shipton derailment (Great Western Railway) of 1874 in which 34 people died. In this case one wheel on the carriage immediately behind the engine broke up following failure of its tyre. The carriage derailed, but was held upright and in line by the pull of the couplings. The driver observed the situation and immediately sounded his alarm whistle and applied full engine braking before the guards at the rear of the train could apply their own brakes (the train was not fitted with continuous brakes). The derailed vehicle was demolished and nine following carriages derailed and toppled over a bridge parapet. In the railway inspector's report it was pointed out that the tyres of the carriage which had suffered the wheel failure were rivetted to the wheel rim, an obsolete form of construction which the Great Western had agreed to discard as dangerous in 1855. Yet the wheels had been re-tyred in the same fashion in 1868. The action of the driver was identified as crucial since had he simply shut off steam and allowed the guards time to apply their brakes it was likely that no serious damage would have resulted. However, the inspector's criticism in this regard was directed at the railway company whose training and rules of operation for engine drivers gave no guidance on how to handle such an event.

In the Shipton example the immediate accident "causes" were mechanical failure and operator error (the same as those most frequently cited for the Three Mile Island accident), but as the inspector's report makes clear, these were the result of significant failures within the operating institution.

In the nuclear power industry, institutional responsibility for safety is recognised in a general sense in the IAEA Safety Guide, *Management of Nuclear Power Plants for Safe Operation*, where it is stated that: "The operating company shall have overall responsibility with respect to the safe operation of its nuclear power plants". More specifically, at a 1983 IAEA safety seminar two senior Canadian nuclear safety specialists, Brown and Meneley, drew attention to the necessity for those at senior levels in an institution to realize: "That they [senior management] must be in control of all safety design and operating decisions throughout the life of the plant. This important task cannot be delegated safely to any other group".

The most recent, and unambiguous, articulation of institutional responsibility for safety was by Lord Marshall of Goring, then chairman of the UK's Central Electricity Generating Board, at the Symposium on Quality in Nuclear Power Plant Operations held in Toronto in 1989, when he noted:

"I must remind you that the ultimate safety responsibility for a nuclear plant cannot rest with any individual engineer because individual people may have been trained badly, or they may be operating a reactor with a design fault or they may simply have been given too much responsibility. The ultimate responsibility cannot rest with the station manager either, because although he has a vital safety role to play, he is implementing the policies of the nuclear utility that employs him. The ultimate responsibility cannot rest with the nuclear designers, because the utility was not compelled to buy a reactor of that particular design. The ultimate responsibility cannot rest with the regulator because if it did, the utility would only need to obey the written regulations. Therefore the ultimate safety responsibility must rest with the corporate organisation that operates the plant. Regulators, designers, manufacturers and individuals all have an important safety role to play, but the ultimate and fundamental safety responsibility must rest with the nuclear utility. Therefore, each nuclear utility has an individual corporate responsibility to guarantee nuclear safety and no amount of international collaboration and discussion can interrupt or replace that responsibility."

The seven reactor accidents discussed here all exhibit to some greater or lesser extent examples of failures to discharge successfully or completely institutional responsibility for safety — that is, they all have elements of "institutional failure".

## Bibliography

M Levenson and F Rahn, *Realistic Estimates of the Consequences of Nuclear Accidents*, Electric Power Research Institute, 1979.

*Nuclear and Non-Nuclear Risk — An Exercise in Comparability*, Report prepared by Pollution Prevention (Consultants) Ltd. for the Commission of the European Communities, EUR 6417 EN, 1980.

L T C Rolt, *Red For Danger: A History of Railway Accidents and Railway Safety*, (Second, extended edition) David and Charles, Newton Abbot, 1966.

*Management of Nuclear Power Plants for Safe Operation*, IAEA Safety Series No. 50-SG-09.

R A Brown, D A Meneley, *Management of Nuclear Power Plant Operation*, IAEA Seminar on Safety in Nuclear Power Plant Operation, Vienna, Austria, November 1983.

# Appendix

## Accidents in nuclear installations which led to death by exposure to ionizing radiation

**08–08–1945** — Los Alamos (USA)  1 dead
Criticality accident: the critical mass was reached by an employee stacking reflector blocks round a subcritical assembly. The employee died. A guard seated 12 meters away received a dose equivalent of 0.32 sieverts.

**21–05–1946** — Los Alamos (USA)  1 dead (20 sieverts)
Criticality accident: during demonstration of a critical mass measurement, a reflector hollow shell was accidentally moved closer to the reactor.

**15–10–1958** — Vinca (Yugoslavia)  1 dead
Criticality experiment without biological shielding: uncontrolled criticality due to an inadvertent rise in the heavy water level (due to an operator error) led to six staff being exposed. Consequence: one dead and five people treated by bone marrow graft in Paris.

**30–12–1958** — Los Alamos (USA)  1 dead (60 sieverts)
Criticality accident: the transfer of a solution containing fissile material from a safe geometry vessel to an unsafe geometry vessel led to the exposure of three staff members (respectively 60, 1.3 and 0.5 sieverts).

**03–01–1961** — Idaho Falls (USA)  3 dead
SL1 research reactor: gross reactivity insertion through manual removal of control rod: two persons killed on the spot by an explosion due to a "water blast". A third person died two hours later (head wound). (See p31).

**24–07–1964** — Woods River Junction (USA)  1 dead
Criticality accident: error during transfer of a highly enriched uranyl nitrate solution to a vessel.

**13–05–1975** — Italy  1 dead
Exposure from a cobalt 60 source in a food sterilization plant (human error).

**23–09–1983** — Constituyentes (Argentina)  1 dead
An accidental power excursion, supposedly due to nonobservance of safety rules during a core modification sequence, resulted in the death of the operator, who was probably only 3 or 4 meters away.
    Assessments of the doses absorbed by the victim range from 5 to 20 Gy for the gamma dose, together with from 14 to 17 Gy for the neutron dose.

The operator dies 48 hours after the accident. Other people in the control room at the time only received very slight doses(1).

## 26–04–1986 — Chernobyl (USSR)   31 dead

Power runaway accident, destroying the reactor and resulting in two immediate deaths from burns and multiple trauma. Most of the radioactive products were discharged to the atmosphere during a 10 day period. Practically 10% of the core inventory at the time of the accident was released: 100% of the rare gases, 15 to 20% of the volatile products, 3 to 5% of the non-volatile products and transuranic elements. About two hundred people suffered acute radiation sickness, twenty-nine of whom died during the three months following the accident. (See p 81).

## Accidents in nuclear power stations which had consequences on the environment, exposure of personnel and plant availability

## 1. Accidents which had consequences on the environment and the public

Among the serious accidents in nuclear power reactors, only one (that of Windscale) is really significant. All the others were material incidents, of which only a few had consequences concerning exposure of personnel.

## 1957 — Windscale (Great Britain)

Reactor fire caused by degraded cooling of uranium metal fuel elements.

Fission products (mainly 740 terabequerels of iodine 131) were released into the environment. Immediate full-scale monitoring of the environment and persons was established. In particular, measures were taken to check and, where necessary, stop milk deliveries by producers in the region.

238 persons were examined. 126 showed signs of slight contamination at the thyroid level (the highest dose was 0.16 sievert). Among the plant workers, 96 were slightly contaminated (0.1 sievert maximum at the thyroid level, despite wearing a mask). As for external exposure, 14 workers received dose equivalents of up to 47 millisieverts. (See p21).

---

(1) The quantity of energy absorbed per mass unit of an irradiated medium is called a "dose", expressed in "gray" (1 gray = 1 joule/kg) or in "rad" (1 gray = 100 rads). The dose equivalent absorbed by a living organism reflects in addition the way in which this energy is distributed in space, depending on the characteristics of the radiation considered and, in the event of internal contamination of bone tissue, the localization of the radiation source. The absorbed dose equivalent is expressed in "sievert", when the dose is expressed in "gray" and in "rem" when the dose in expressed in "rad" (1 sievert = 100 rem).

## 2. Accidents which led to staff being exposed above the permissible level

**1958** — Chalk River N.R.U. (Canada)

During the unloading of a defective fuel element in this experimental reactor, the element became stuck in the transport container and part of it fell into the storage pit where it burned. The accident and the subsequent repair work led to three people receiving 100 to 200 millisieverts, 15 people 50 to 100 millisieverts, 30 people 30 to 50 millisieverts, and 104 people 10 to 30 millisieverts.

A very small amount of radioactive products was released into the environment over an area of 40 hectares (about 100 acres) which was practically uninhabited.

**March 1965** — Chinon A1 (France)

A worker who passed a no-entry beacon and entered the fuel unloading premises received a dose of 0.50 gray.

**September 1979** — Chinon A2 (France)

During a search for the source of a leak of carbon dioxide — the reactor's main coolant — two workers received dose equivalents of 340 and 110 millisieverts.

## 3. Accidents which had consequences on the availability of plants

HEAVY WATER REACTORS

**1952** — NRX

Extensive fuel melting following a power excursion. (See p13).

**1969** — Lucens (Switzerland)

Destruction of a pressure tube and consequential failure of adjacent pressure tube, partial core melt, high contamination of the vault containing the reactor, but no significant activity outside. The plant was decommissioned. (See p53).

**1968** — EL4 (France)

The replacement of faulty steam generators led to the shutdown of the plant for two years.

GAS-COOLED REACTORS

**1967** — Chapel Cross (Great Britain)

Subsequent to melting of fuel elements, the reactor remained shut down for two years.

**1969** — Saint-Laurent A1 (France)

Subsequent to the melting of five fuel elements, 50 kg of uranium were dispersed in the reactor vessel, and the plant was shut down for a year.

**1980** — Saint-Laurent A2 (France)

Subsequent to the obstruction of 6 to 8 graphite stack channels by a metal plate, 2 fuel elements melted. The reactor was shut down for around two and half years.

## PRESSURIZED WATER REACTORS

— Incidents which took place in the reactor internals (breaking of bolts and centering parts, movement of the thermal shield) led to plants being shutdown for relatively long periods:

> **1967** — Trino Vercellese (261 MWe) — Italy — 3 years.
> **1968** — Sena-Chooz (300 MWe) — France — 2 years.
> **1969** — Novoronezh (196 MWe) — USSR — 18 months.
> **1972** — Oconee (887 MWe) — USA — 8 months.
> **1973** — Palisades (668 MWe) — USA — 1 year.
> **1976** — Biblis (1,146 MWe) — FRG — 4 months.
> **1979** — Three Mile Island 2 (900 MWe) — USA
> At TMI faulty sealing of the pressurizer relief valve and the premature shut-down by operators of the safety injection and of the primary pumps, following a pressure transient provoked by a loss of steam generator feedwater supply, led to partial melting of the core and extensive contamination inside the containment. Radioactive release into the environment was very limited. (See p69).
> **1982** — Gravelines 1, Fessenheim 1, Bugey 2 and 4 (900 MWe) — France
> The support pins for the guide tubes or rod cluster control assemblies broke and led to these plants being unavailable for several months.
> **1985** — Paluel 1 and 2, St. Alban 1, Flamanville 1, Paluel 3 (1,300 MWe) — France
> Leaks on the in-core instrumentation thimble guide tubes led to unavailability of the first two units and delayed initial startup of the other three by a few weeks.

— Incidents (tube rupture, loss of feedwater etc.) occurred in the steam generators of the following power stations:

Beznau 1 (350 MWe) — Switzerland.
Indian Point (225 MWe) — USA.
Mihama 1 (320 MWe) — Japan.
Robinson 2 (740 MWe) — USA.
Surry 1 and 2 (800 MWe) — USA.

Oconee (886 MWe) — USA.
San Onofre (450 MWe) — USA.
Turkey Point 3 and 4 (700 MWe) — USA.
Palisades (800 MWe) — USA.
Ginna (490 MWe) — USA.
Davis Besse (934 MWe) — USA.

— Incidents (cladding deterioration, densification of fissile material) took place on the reactor core fuel assemblies in the following power stations:

Ginna (490 MWe) — USA.
Beznau 1 (350 MWe) — Switzerland.
Surry (800 MWe) — USA.
Zorita (156 MWe) — Spain.
San Onofre (450 MWe) — USA.
Point Beach (500 MWe) — USA.
Robinson 2 (740 MWe) — USA.
Turkey Point (700 MWe) — USA.
Bugey 2 and 3 (900 MWe) — France.

BOILING WATER REACTORS

**1975** — Browns Ferry (2 x 1,065 MWe) — USA.
A fire which was due to inadequate methods for checking the integrity of the sealing of cable penetrations in the containment led to a 17-month shutdown.

**1972** — Millstone 1 (650 MWe) — USA.
A 6-month shutdown was caused by seawater entering the steam circuit due to a leak in condenser tubes.

**1972** — Wurgassen (640 MWe) — FRG.
An incident in the pressure suppression system in the containment led to a 7-month shutdown.

**1973** — Wurgassen (640 MWe) — FRG.
A leak in one of the main steam lines led to a 5-month shutdown.

FAST BREEDER REACTORS

**1955** — EBR 1 — USA.
Fuel melting led to very slight contamination of the building and decommissioning of the reactor.

**1966** — Enrico Fermi (51 MWe) — USA.

After partial melting of two fuel subassemblies, it took 4 years to get the installation back to normal operation. It was decommissioned in 1972. (See p45).

**1971** — KNK (20 MWe) — FRG.

A sodium fire in the reactor building following a large leak (between 500 and 1,000 kg of sodium) led to a 4-month shutdown.

**1973** — BN 350 (350 MWe) — USSR.

A sodium-water reaction took place in the steam generators.

**1976** — Phenix (250 MWe) — France.

A secondary sodium leak at two intermediate heat exchangers led to a 15-month shutdown.

**1982** — Rapsodie — France.

A leak was detected in the two vessels around the reactor.

**1982–1983** — Phenix (250 MWe) — France.

Three steam generators were affected by a sodium-water reaction (in April and December 1982 and February 1983), without provoking a long-term shutdown of the plant.

# NRX

The NRX (National Research Experimental) reactor was first started up in 1947 at the Chalk River Nuclear Laboratories in Canada. A heavy water moderated, light water cooled research and plutonium production reactor, NRX had an initial maximum power of 20MWt. By 1952 power had been uprated to 30MWt.

The reactor essentially comprises a 103.048m high by 2.743m diameter vertical aluminium cylinder (the calandria) containing the heavy-water moderator, threaded by about 200 vertical tubes accommodating the fuel/coolant assemblies

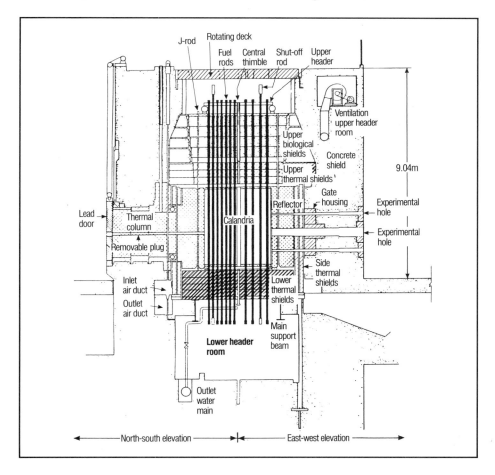

*Cross-section through the NRX reactor.*

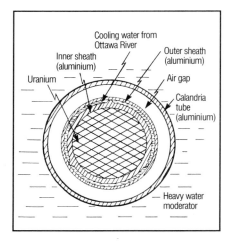

Cooling water from Ottawa River

Inner sheath (aluminium)

Uranium

Outer sheath (aluminium)

Air gap

Calandria tube (aluminium)

Heavy water moderator

*Cross section through the NRX fuel/coolant assembly.*

and the shutoff/control rods. The calandria is surrounded by a graphite reflector, concrete biological shielding and steel and concrete upper and lower thermal and biological shields. An inert atmosphere is maintained over the moderator by means of a helium cover gas system.

The fuel/coolant assemblies in use in 1952 principally consisted of solid natural uranium metal rods clad with aluminium. The cladding is provided with three ribs which support an outer cooling jacket. Cooling water from the Ottawa river is supplied to an upper ring header at 175 lb/in$^2$ and 10°C, flows down the three annular spaces in each fuel assembly to a lower (outlet) ring header, exiting at a temperature of 50°C. After hold up in delay tanks to allow any short-lived radioactivity to decay the water is returned to the Ottawa River.

Reactor control was via 12 pneumatically driven boron carbide rods, variation in moderator level (a 3cm change in moderator level was worth 1 mk) and a single cadmium rod (worth about 9 mk) for fine control. Reactor shutdown could also be accomplished through moderator draining — 30 seconds being required to drain from full to half level. Each shutoff rod included a 43.18cm long drive piston and weighed a total of 13.2kg — the rods were designed to be as light as possible to permit rapid acceleration and deceleration. When driven by 100 lb/in$^2$ compressed air the rods would fully insert within one second; falling free under gravity, insertion time was three to five seconds. The fully "up" position of each rod was indicated by a lamp on the control panel and the rods were held in the fully raised position by electromagnets.

The 12 rods were divided into six groups or "banks" of 4, 3, 2, 1, 1 and 1 rods respectively, the first bank of four rods being designated the "safeguard" bank. Electrical interlocks using limit switches at the bottom of the rod travel were designed to preclude the withdrawal of any rod before withdrawal of the "safeguard" bank. The rationale for this arrangement has been outlined as follows [Lewis, 1953]:

> "The design reason for distinguishing the safeguard bank is that, for safety, no shutoff rod may be raised unless either (a) more than seven shutoff rods would be left fully down or (b) more rods are available for quick release than are being raised at any time. To make startup possible, some rods must satisfy condition (a) and not (b), and, if the total of shutoff rods is only 12, no more than four may be set for condition (a). All other rods

must satisfy condition (b). To achieve a safe startup in the shortest time, as large a number as possible and the most highly effective rods were in the safeguard bank."

The rods were controlled by four pushbuttons, of which three (#1, #2 and #3) were mounted on the control desk and the fourth (#4) on a wall panel to the left of the desk. Pushbutton #1 raised the four-rod "safeguard" bank. Pushbutton #2 raised the remaining banks in sequence. Pushbutton #3 increased the electric current to all the solenoids in the shutoff rod headgear system to seat the air valves firmly (to preclude subsequent leakage of $100lb/in^2$ air) and to draw the shutoff rods fully home. Pushbutton #4 charged the rod headgear with compressed air. The thinking behind the location of pushbutton #4 was that since pushbuttons #3 and #4 had to be pressed at the same time so that the high pressure air would not leak past inadequately seated valves, the locations of #3 and #4 should be selected to emphasise the special nature of their operation.

The usual startup procedure for NRX was:

❏ Raise the moderator level to one somewhat lower than that predicted for criticality.
❏ Withdraw all the shutoff rods.
❏ Fully raise the single control rod.
❏ Slowly raise the moderator level until criticality is reached.

## Accident sequence

A low power experiment had been scheduled to compare the reactivity of long-ir-radiated fuel with that of fresh fuel. Ten fuel rods were operating with reduced coolant flows and one rod was air-cooled. The ten reduced flow rods were fed either by hose connections or temporary "jumper" connections to other rods — these provisions being made to facilitate a change from water to air cooling in the course of the experiment.

The interlocks to preclude withdrawal of any rod before the "safeguard" bank were out of service on 12 December owing to defects in the limit switches. In order to obtain the flux distribution required for the experiment six shut-off rods were to remain in the core and appropriate adjustments had been made to the valve settings in the rods' air supply system. Howeyer, it was discovered that even with the reactor loaded with these six rods, moderator height was still lower than the desired level so it was decided that a seventh rod should be added to the "load". It was during the adjustments to the air valves necessary to accomplish this that an operator in the basement opened by mistake three or four bypass valves on the shutoff rod air system, causing three or four rods to rise fully out of the core. The supervisor at the control desk saw the panel lamps indicating rod withdrawal and immediately went down to the basement and closed the bypass valves. The rods should have dropped fully back into the core and the supervisor assumed that they had indeed done so, but in fact at least two or three of the rods did not drop back in, although they did fall far enough to clear their top-of-travel contacts and allow the panel lamps to go out. The supervisor then telephoned his assistant at the control desk and told him to press pushbuttons #4 and #1. The supervisor had intended to say #4 and #3, and attempted to correct the instruction but since his assistant had put down the telephone in order to operate the two pushbuttons he could not be recalled. The compressed air supplied to the rod headgear leaked out past the valves (pushbutton #3 not pressed) and the four rods of the "safeguard" bank rose out of the core.

As was mentioned earlier, it had been assumed that the rods originally raised in error had dropped back into the core, so even with the safeguard bank raised NRX should still have remained sub-critical. The fact that it did not was a considerable surprise to everyone in the control room.

Power rose to about 100kWt over a 20 second period, at which time the reactor was tripped. Even without the benefit of compressed air, the rods should have fallen fully back into the core within three to five seconds. In fact, it was subsequently determined, only one rod did fall, very slowly, taking about 90 seconds to enter fully. Power continued to rise and had reached 17MWt at 30 seconds at which point reactivity suddenly increased by about 2.5 mk as the low coolant-flow rods boiled dry. Continuing efforts to drive the shutoff rods down proving unavailing, moderator dumping was begun at 45 seconds.

An estimated peak power of about 90MWt was reached at 49 seconds but within 20 seconds the moderator dumping had taken full effect and the reactor was back at low power.

About three minutes later a "rumble" was heard, accompanied by a spurt of water through the top of the reactor, and the helium gasholder abruptly rose to the top of its travel — a phenomenon which has been attributed to a hydrogen burn inside the calandria. At about this time evacuation of the plant site was ordered since activity levels in buildings with forced ventilation had become higher than external levels.

The highest personal exposure outside the NRX building at the time of the accident was 350mRem, received by an electrician who was up a pole adjacent to the reactor stack.

The reactor was extensively damaged with fuel melting taking place in 22 locations, including all the low coolant flow rods and the air-cooled rod. Heavy radioactive contamination (especially in the lower header room) considerably hampered recovery and repair work. Nevertheless, in an unprecedented operation, the damaged calandria was removed and replaced and the reactor rebuilt and returned to service within 14 months. The maximum radiation dose to work crews during clean-up and calandria removal was 17 Rem and the majority of the workforce received less than 3.9 Rem.

In 1970 the NRX calandria was replaced a second time and the reactor remained in service until 1987, since when it has been maintained in "hot standby" condition.

## Discussion

The single most important technical lesson of the NRX accident is that a reactor should always have fast shutdown capacity available and that this capacity should be independent of any control system. Following the accident, Lewis and Ward examined the problem of safety shutdown in some detail. As they pointed out, contemporary thinking was that the safest reactor condition was that with every available piece of neutron absorbing material in the core. In fact, and perhaps counter-intuitively, this is not the case:

> "... in any operation it is possible that the reactivity of the pile will be changed. In many cases it is knowingly changed, for example by the removal of a shutoff rod for maintenance, by altering the pile load or fuel or moderator. Also, however, equipment failures or errors can change the pile reactivity. In the normal shutdown state if all the shutoff rods are in, then the monitoring instruments have no control and cannot guard against the unintended reactivity changes."

This shutdown capacity must always be available and operational reactivity manoeuvres must be firmly separated from the ability to insert swiftly sufficient negative reactivity into the core to achieve unequivocal shutdown. In the case of the NRX accident, in effect seven out of 12 rods in the shut-down system had been temporarily converted to use as a control system. In the process of carrying out the conversion, errors in manipulation of the air valves set in motion the sequence of events which led to serious reactor damage. Simply put, the only role for a safety shutdown system should be fast shutdown when invoked by specified reactor conditions.

Lewis and Ward also outlined some of the conditions under which fast shutdown should be invoked, noting the balance that must be achieved between an "over sensitive" system causing frequent and "spurious" trips, which can unduly impede operation and give rise to operator frustration and possible interference with the system, and a system which is sensitive and fast acting enough to achieve shutdown before fuel damage occurs. The classical way of limiting the behaviour of any system is to define the upper limits of operation and to set a detector close to that limit to order and effect the limitation or termination of operation.

However, this "overpower" approach, while simple and reliable, is of maximum efficacy when the machine is operating at or close to its full rated power. Consider a nuclear fission reactor operating at low power with its shutdown system sensitized only to over (maximum) power. A sudden insertion of reactivity (say more than 2 mk) could bring about a situation in which power is increasing at a rate that may seriously challenge the speed of operation of the shutdown system by the time maximum power is reached. Lewis and Ward concluded:

> "For this reason it is very advisable to have prior warning of such behaviour of the reactor, and this can be provided by trips actuated by measurements of d/dt and d(log)P/dt where P is reactor power".

The importance of the NRX accident in focussing attention on the reactor shutdown system, and the resultant evolution of a design philsophy for this most vital of reactor safety systems cannot be overstated. However, consideration of the events leading up to the accident does reveal influences on reactor safety other than engineering design and direct operator action. Lewis [1953] points out that: "To reduce the risk of human error and mechanical failure, no doubt a better system of review and inspection should be established. This should relate the design considerations to the current experience".  Had NRX operations been supported with the appropriate structures and procedures for review of operations and proposed experiments then it is plausible to suggest that:

❏   Operation of the reactor in a degraded condition, with the shut-off rod removal interlocks unavailable, might have been precluded. It should be noted that under normal operating conditions, withdrawal of any shut-off rod/s before safeguard bank withdrawal would not pose a hazard since criticality would not be approached until all rods (including the single control rod) had been withdrawn. However, under the conditions of the December experiment, with seven of 12 rods adjusted to remain in the core, the margin was much narrower and the importance of the interlocks correspondingly greater.

❏   Formal written procedures for operation of the shut-off rod system's air control valves would have been available. In addition, the air bypass valves would have been clearly tagged for non-operation (Lewis [1953] notes that the handles on some of these valves "had been removed for safety").

❏   Greater consideration would have been given to the reactivity implications of operation with a symmetrical concentration of fuel rods with very low coolant flows in a central region of the core. Five of the rods with low coolant flows were concentrated in Circle No 2, in the positions L-15, K-18, H-18, H-12 and K-12, and were fed via hose connections with about 1 gpm of water. The only other rod in this circle, G-15, was air cooled. The effect of the low flow rods boiling dry was to insert signficant reactivity (about 2.5 mk) at a significant stage in the sequence of events. Lewis notes that had this reactivity addition not taken place, then power would have levelled off at about the 20MWt level as the single shut-off rod dropped slowly back into the core.

# Bibliography

W B Lewis, *An Accident to the NRX Reactor on December 12, 1952,* Atomic Energy of Canada Ltd, AECL-232, 13 July 1953.

W B Lewis, Letter to T J Thompson, 13 April 1964.

D G Hurst, *The Accident to the NRX Reactor, Part II,* Atomic Energy of Canada Ltd, AECL-233, 23 December 1953.

J W Logie, *Three Vessel Replacements at Chalk River,* AECL-6899, Atomic Energy of Canada Ltd, 1980

W B Lewis and A G Ward, *An Appreciation of the Problems of Reactor Shut-off Rods with Special Reference to the NRX Reactor,* AECL-490, Atomic Energy of Canada, May 1953

# WINDSCALE

The Windscale Piles Nos 1 and 2, completed in late 1950, were plutonium production reactors constructed as part of Britain's nuclear weapons programme.

Graphite moderated and forced air-cooled, the reactors were originally designed to use natural uranium metal fuel. However, after a short period of operation very slightly enriched uranium (0.73 per cent U-235) was introduced in 1953. The level of enrichment was chosen to give the minimum production cost for plutonium. Gowing notes that this move meant that the piles' "heat rating went well above the design figures".

The reactor core essentially comprised a rectangular cross-section stack of graphite blocks pierced by horizontal fuel channels on a 20.95cm square lattice. For fuel loading the channels were arranged in groups of four, access to each group being via a single charge hole in the front shield. In the centre of each group of four channels a single, smaller diameter channel was used for production of other isotopes. Control was via horizontal absorber rods and vertical shutdown rods. Additional vertical channels through the core were provided for experiments in support of the civil nuclear power programme. The uranium metal fuel was in the

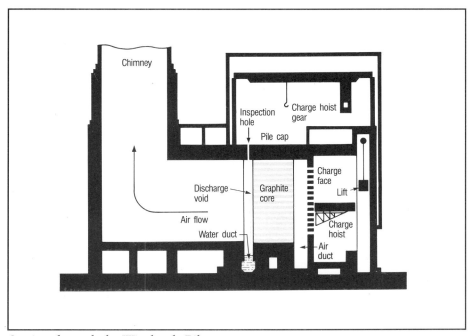

*Section through the Windscale Pile.*

form of short rods sealed in finned aluminium cans resting in graphite "boats". Following irradiation, fuel rods would be pushed through the rear (discharge) face of the reactor to fall into a water-filled trench from which they would be removed by skips to cooling ponds where, after removal of the graphite boats, they would await processing. The cooling circuit was a once-through forced air system, the outlet air being discharged through a 137.16m stack with filters at its top.

## Background to the accident

At the time the Windscale piles were designed knowledge of the effects of radiation on graphite was limited. It was understood that graphite expanded under irradiation but, in 1949, data on the amount of growth was conflicting. When the Windscale Piles were started up, weekly measurements of graphite growth were taken, but no operational plans had been made to deal with the resulting stored energy (called Wigner energy*).

In September 1952 a spontaneous release of Wigner energy occurred while the pile was shut down. As a result of this a method was instituted for the controlled release of Wigner energy at regular intervals and thermocouples were installed to monitor graphite temperature.

The method adopted to release the energy consisted of shutting off the cooling air flow, bringing the reactor to criticality and allowing the graphite temperature to rise until it had reached the point at which the Wigner release started. The reactor was then shut down again.

Up to the end of 1956 eight such operations had been carried out on Pile No 1, but not always successfully. An April 1956 attempt failed completely and two others achieved only partial success with energy releases being recorded from only some regions of the core and "pockets" of graphite remaining unannealed.

Wigner release intervals were originally set at 20 000MWd, subsequently extended to 30 000MWd. In view of what seemed to be the increasing difficulty of obtaining successful releases, in 1957 consideration was being given to a significant further increase to 50 000MWd intervals. However, until more experience had been gained a 40 000MWd interval was selected, implying a Wigner release operation for October of that year.

---

* Below a temperature of about 350°C, at which annealing of defects cannot take place, radiation damage increases with time, causing the accumulation of energy in the crystal lattice of the graphite. If this metastable material, loaded with stored energy, is suddenly transformed to the stable form, the excess stored energy can be released at once, causing a large increase in temperature. Graphite moderated reactors operating at fairly low temperatures follow procedures to allow the controlled and gradual heating of the material so that annealing of radiation damage can take place. At high operating temperatures the problem does not arise since the annealing and heat release proceed continuously. [F J Rahn *et al*, *A Guide to Nuclear Power Technology*, Wiley-Interscience, 1984, p441].

## The accident sequence

On 7 October 1957 Pile No 1 was shut down at just after 01:00 in preparation for the planned Wigner release operation. Preparatory work included checking (and, where necessary, replacing) the uranium and graphite thermocouples used for following the Wigner release, switching off the shutdown cooling fans and opening roof inspection hole covers and a door in the base of the stack to minimise cooling air flow through the pile.

About 18 hours after shutdown (19:25) the reactor was made critical and the power level gradually raised to an indicated 1.8MW by 01:00 on 8 October.

This would be a low reading since with the upper horizontal control rods fully inserted (as was the case for this operation) the ion chamber measuring reactor power would be masked. However, since for Wigner release operations power was controlled on the basis of temperature readings, the only effect of this masking would be to alter the relation between power changes and movement of the lower horizontal control rods.

At about this time a fuel cartridge temperature of 250°C was indicated in two channels and since this was laid down as an upper limit for the initial stages of a Wigner release the control rods were run in and the reactor shut down by 04:00.

Initially, graphite temperatures rose in the manner characteristic of a Wigner release, but by about 09:00 (8 October) the physicist in charge felt that the graphite temperatures seemed to be falling rather than rising and further heating was needed if the release was to be completed.

Second nuclear heatings had been applied during three previous Wigner energy releases in 1954, 1955 and 1956, but had not been applied until 24 hours after the last temperature rise and when all graphite temperatures were observed to be falling. In this case, however, the second heating was initiated at 11:05. Uranium temperatures rose sharply, the highest thermocouple indication being 380°C about 15 minutes after restart. Power was reduced and the thermocouple reading fell to 334°C within 10 minutes. The pile was maintained at this lower power level until 17:00 at which time it was shut down.

In the course of the following day (9 October) graphite temperatures showed considerable variation, but with a general tendency to increase. For example, at one location (channel group 20/53) the temperature rose steadily from an indicated 255°C at the time of the start of the second heating to an indicated 405°C at 22:00, some 36 hours later.

Because of this high temperature trend the chimney base door and inspection holes were closed at 21:00 to allow natural chimney draught to induce a cooling flow of air through the pile. However, the cooling effect was not considered adequate and at 22:15 the fan dampers were opened for 15 minutes to provide a positive flow of air. This operation was repeated at 00:01, 10 October (10 minutes), 02:15 (13 minutes) and 05:10 (30 minutes) and resulted in a fall in all graphite temperatures save the highest (20/53), where the temperature rise was merely halted.

At the end of the fourth damper opening (05:40) the stack activity monitor showed a sharp increase, which was attributed to the first movement of air through the pile. Over the next two and a half hours this reading fell steadily, but then began to rise again. The graphite temperature in channel group 20/53 continued to rise, and at 12:00 a temperature of 428°C was recorded, prompting further damper openings at 12:10 (15 minutes) and 13:40 (5 minutes). During these openings a second, and very much larger, increase in stack activity was recorded and at about the same time high activity readings on a nearby roof (the Meteorological Station) were reported. From this it was inferred that one or more fuel failures had occurred and at 13:45 the shutdown cooling fans were switched on to drive air through the pile so that the channel scanning equipment could be used to identify the location of the failed fuel. The scanner, however, proved to be jammed - a situation experienced during previous Wigner energy releases, and probably attributable to overheating.

Sampling of the exhaust air from the pile revealed high levels of particulate activity and suggested that serious fuel failure must have occurred, necessitating rapid identification of the affected channel and discharge of the fuel. Channel group 21/53 had shown very rapid temperature increases and, by 16:30, had an indicated temperature of about 450°C. Since the scanning gear was not working this channel was inspected visually and the fuel inside was seen to be glowing. Initial attempts to discharge the fuel were fruitless and consideration was given to blanking off the channel group using graphite plugs.

However, shortly afterwards (about 17:00) it was found that around the affected group of channels a total of about 150 channels (40 groups) were at red heat. The graphite was effectively on fire.

Two expedients were then adopted: discharge of fuel from the hot channels; and discharge of fuel from a ring of channels surrounding the hot channels in order to create a firebreak. This latter was successfully accomplished and two more rows of channels above the hot region were subsequently discharged as the fire threatened to spread upwards.

Attempts were made to extinguish the fire using $CO_2$ but these were unsuccessful, and successive observations through an inspection hole revealed a steady growth in the fire. A glow at the rear face of the pile was evident at 18:45. At 20:00, yellow flames were seen and half an hour later the flames were blue.

At this point the use of water was first considered, even though it presented two potential hazards. First, the danger of a hydrogen-oxygen explosion which could blow out the stack filters and lead to a large and uncontrolled release of radioactivity; and second, a possible criticality hazard through the introduction of water into an air-cooled pile.

These dangers, however, had to be balanced against the distinct possibility of the whole pile catching fire should the graphite temperature rise much higher than 1200°C.

By midnight it was decided that if all other measures failed to achieve a temperature reduction, then water should be used. The fire brigade was ordered to standby and arrangements were made to enable water to be injected into the discharged channels.

At 01:38 the graphite in channel group 20/53, near the top of the high temperature area, had an indicated temperature of 1000°C and optical pyrometer readings in this area indicated a fuel temperature of about 1300°C.

Strenuous efforts over the next two hours were successful in dislodging burning fuel elements from the top row of channels in the "hot" area, but by this time the fire was too widespread for such actions to have very much effect.

At 04:00 blue flames were still in evidence and the graphite appeared to be burning. Efforts to discharge the burning fuel continued, but by 07:00 it was clear that the fire was not being checked and it was decided to flood the pile. After all the site workers had taken cover, water injection started at 08:55.

For the first hour there was little observable effect and flames were stll visible, but after the shutdown cooling fans were turned off at 10:10 the fire imediately began to subside. Water injection continued until 15:10 on 12 October by which time the pile was cold.

## Immediate causes of the accident

In their report to the Chairman of the United Kingdom Atomic Energy Authority, the Committee of Inquiry chaired by Sir William Penny concluded that the fire began during the second nuclear heating which was applied too soon and too rapidly. Most probably the rapid rise of temperature during the second nuclear heating caused the failure of one or more of the aluminium fuel element cans. The exposed uranium oxidised, providing an additional heat source which, in combination with the heating effect of the later Wigner releases, started the fire.

The Report drew attention to the fact that, contrary to the pile physicist's impression that the intial nuclear heating had not triggered a Wigner release, examination of the thermocouple traces showed that a "substantial number" of graphite thermocouple readings showed steadily inreasing temperatures. In previous Wigner release operations where application of a second nuclear heating had been made, this had not begun until virtually all graphite temperatures were observed to be falling.

The second heating was also applied at an unusually rapid rate — the recorded rate of uranium temperature rise was 10°C/min, in contrast with the 2°C/min rate adhered to under normal operational practice. In addition, the positioning of thermocouples in the pile was such that misleading uranium temperature indications would be given. Since in Wigner release operations the control rods were adjusted to give a flux peak 0.9m closer to the front of the pile than under normal operation and airflow through the pile was minimised, maximum uranium tem-

"*A helicopter hovers over the 445ft chimney of the No 1 pile at the Windscale plutonium factory, so that a scientist can make tests from a safe distance for the inquiry into the recent overheating of the uranium charges. It was considered too dangerous to use the lift built on the outside of the chimney, which is topped by a filter which prevented all but a fraction of the deadly radio-active particles from escaping.*" This was the caption to an official photograph put out on 20 October 1957.

peratures would occur about 2.1m closer to the front of the pile than the thermo-couples. This would result in maximum uranium temperatures being about 40 per cent higher than the thermocouple readings.

Calculations suggested that, at the position of peak neutron flux, uranium temperature would have risen rapidly from an initial value of 340°C to as high as 450°C in the course of the second nuclear heating, and would have remained at this level for some minutes. Based on the known behaviour of the Windscale fuel elements and their irradiation history, the Report concluded that this treatment would have brought about immediate clad failure for fuel in this region.

That the fuel failures were not diagnosed earlier is attributed to the nature of the airflow through the pile. The initial attempt to induce cooling air flow through the pile at 21:00 on 9 October involved closing the chimney base door and the inspection holes and opening the inlet air dampers. At this point the air in the pile would be stagnant and that in the chimney would be cool, so there would be little differential pressure available to drive air through the core and carry fission products up the stack to the monitors.

The subsequent second and third damper openings were too short in duration to establish air flow. But at 05:10 on 10 October the fourth opening of the dampers for a 30 minute period was sufficient to drive gaseous fission products and particulates up the chimney, giving rise to the sharp increase in the stack monitor reading at 05:40. However, at the time this was attributed to the first arrival at the monitors of air which had been resident in the core for some time and hence would carry a larger concentration of radioactive material.

The operations to establish air flow through the core naturally caused an accelerated rate of oxidation of the uranium fuel, while switching on the shutdown cooling fans at 13:45 (following the very large stack monitor readings noted at 05:40) caused a further rapid increase so that by 15:00 an intense fire was burning in the region of the 20/53 channel group.

The initiating events of the Windscale fire could be described as "operator error" on the part of the pile physicist, exacerbated by inadequate instrumentation of the pile. The thermocouples were not correctly positioned for monitoring maximum uranium temperatures during a Wigner energy release and the conditions in the core (ie stagnant air) precluded early detection of the damaged fuel elements. However, the Official Report was at pains to point out that there appeared to exist no form of operating manual for the pile physicist's guidance. The only piece of documentation available appeared to be the following memo-randum dated 14 November 1955:

"Will you please issue the following operating instructions to the operator enaging in controling the Wigner Energy Release. If the highest Uranium or Graphite temperature reaches 360°C, then Mr Fair, Mr Gausden and Mr Robertson are to be informed at once, and the PCE alerted, to be ready to insert plugs and close the chimney base. When the maximum operating temperature reaches 380°C unless further instructions to the contrary have been received the roof plugs will be inserted and the chimney base closed. At 400°C all of the dampers in the blower houses are to be opened and at 415°C shutdown fans are to be started up".

All other details of pile operation existed either as committee minutes or seemed to be traditions (habitual practice) without any form of written authority at all. This lack of formal documentation was identified as a "serious defect".

## Discussion

The Report on the Windscale accident is quite explicit in its insistence that the "causes" of the Windscale fire lie deeper than operator error exacerbated by design inadequacy. The lack of formal operating documentation is to an extent understandable within the context of Windscale operations as a whole, which formed part of a pioneering and urgent programme. Those production pressures which must necessarily exist under such circumstances, as well as the novelty of the process, would naturally tend to encourage a situation where practices evolve on the basis of operating experience and are most efficiently and rapidly promulgated through informal oral communication.

However, another factor, quite carefully identified in the Windscale Report, was a set of "deficiencies and inadequacies of organization". Unclear division of responsibilities between the various branches of the Industrial Group and the technical advisers at Harwell led to failures in communication. In some cases this meant that one group within the AEA would either remain unaware of technical changes made by another or unaware of the significance of those changes for their own work.

It was, the report noted, uncertain "who is responsible for particular technical decisions" and there appeared to be "undue reliance on technical direction by committee". This last point appears to prefigure the comment by Thompson about the organizational deficiencies related to the SL-1 incident, in that a single body must have responsibility for safety and that "a line organisation" is needed, "not a committee".

It was also noted that the operations staff at Windscale were "not well supported in all respects" by technical advice, a crucial example being the fact that the Windscale Works Technical Committee had never examined the records of the recent Wigner releases. This might be seen as another symptom of the sense of urgency, or "production imperative" that undoubtedly existed at the Windscale site. Indeed, the report points to other indications of this, including changes in operating procedures "with a general tendency to push pile temperatures upwards without complete realization of all the technical factors involved."

Specific attention is drawn by the report to the fact that following the start of operation of the Windscale reactors, "other demands on the [UKAEA] Industrial Group have been so heavy that insufficient technical attention has been available to ensure [their] safe operation" and firm recommendations were made for the strengthening of the Windscale organization as well as for a clearer definition of the responsibilities of the several senior safety staff.

## Bibliography

M Gowing, *Independence and Deterrence: Britain and Atomic Energy 1945–1952,* Volume 2, *Policy Execution,* Macmillan, 1974.

W Penney, B Scholand, J Kay and J Diamond, *Report on the Accident at Windscale No 1 Pile on 10th October 1957,* Report to the Chairman of the United Kingdom Atomic Energy Authority, October 1957.

# SL-1

The SL-1 (Stationary, Low-power) reactor was a 3MWt prototype natural circulation boiling water reactor designed for meeting the electrical power and space heating requirements of remote military installations.

The reactor was designed by Argonne National Laboratories (ANL) to meet the following requirements:

❏ The reactor components had to be transportable by air.
❏ The reactor would have to operate for three years on one core loading.
❏ Water requirements should be low.
❏ Above-ground construction should be used.

Located at the National Reactor Testing Station, Idaho, the reactor achieved criticality on 11 August 1958 and, under the direction of ANL, testing (including a 500 hour full-power run) continued until 5 February 1959 when the reactor was handed over to Combustion Engineering (C-E) for operation as a test, demonstration and training facility.

The reactor was housed in a 11.6m diameter by 14.6m high cylindrical building made of 0.63cm steel plate. The lower part of the building was occupied by the reactor vessel itself, surrounded by local gravel for shielding purposes. The middle third, or operating level, of the building contained the turbine generator, feedwater

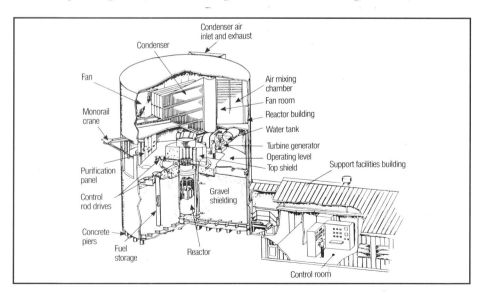

*Cutaway view of the SL-1 plant.*

pumps and other associated equipment. The top plate of the reactor and its protruding control rod housings was surrounded and covered by removeable shielding blocks which were penetrated by the horizontal drive shafts for the control rods. The upper third of the building (the fan room) housed the air-to-water condenser and associated fans and ducting.

The reactor vessel was a 4.42m high by 1.37m diameter cylinder of 1.90cm thick steel clad with Type 304 stainless steel. It had a design pressure of 400lb/in$^2$. Operating pressure was 300lb/in$^2$ with a steam outlet temperature of 215.5°C.

The 20.32cm thick top plate of the reactor vessel was penetrated by nine 15.24cm diameter flanged nozzle openings to accommodate control rods and their shield plugs. The reactor core was a 80.01cm diameter by 65.53cm high cylinder subdivided into 16 boxes, of which 12 could accommodate four fuel assemblies

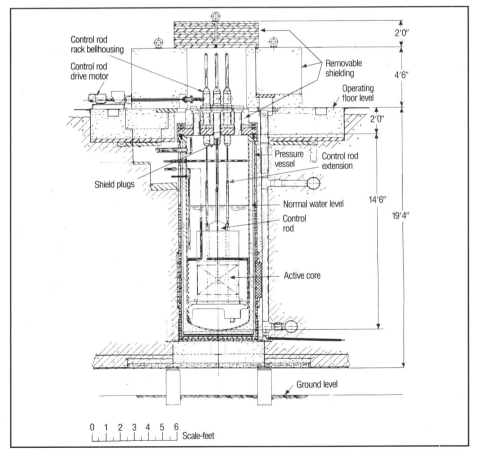

*Cross-section through the SL-1 reactor.*

each and four (the corner boxes) three assemblies. The sides of the boxes were defined by five cruciform and four "T" section control rod channels. Since the initial fuel loading for SL-1 comprised only 40 fuel assemblies, only the five cruciform channels were in use.

Each fuel assembly consisted of nine 0.30cm thick flanged fuel plates spot welded to side plates to form a 9.84cm square by 90.17cm long assembly. The fuel itself was an aluminium nickel uranium alloy with the uranium enriched to 91 per cent U-235, clad with aluminium nickel alloy (X-8001), the first power reactor application of this material. The total core load of U-235 was 14.0kg.

To meet the three year core life criterion, burnable poison in the form of boron-10 was used, with boron powder enclosed in a uranium nickel strip spot welded to one side of each fuel assembly side plate. In the case of the 16 central fuel assemblies a second, half-width, strip was attached to the opposite side plate. The total boron load was 23g.

Control was via five cruciform X-8001-clad cadmium control rods. These were connected to extension rods which, in turn, were connected to rack gears enclosed within bell-housings on the reactor top plate. Each rack gear was driven by a pinion

*The SL-1 cruciform control rod. Note that a 19in (48.3cm) follower was specified for the central control rod - No 9.*

attached to a shaft which, via an electromagnetic clutch, was connected to a drive motor outside the shielding blocks. With the clutch disengaged the rods would fully insert (76.2cm travel) under gravity in two seconds. In the event of sticking the rods could be driven fully down by the motor through a unidirectional cam clutch.

One significant feature of this arrangement was the fact that with the original fuel loading and five control rod configuration, withdrawal of the central control rod (position No 9) was sufficient to achieve criticality.

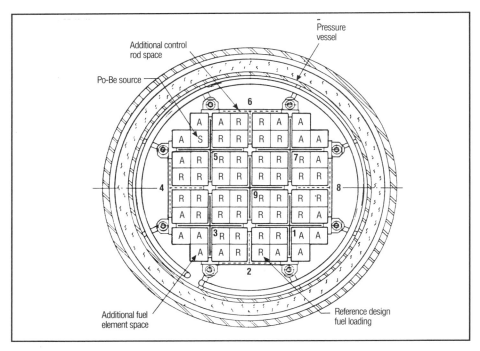

*Plan view of the SL-1 core. The numbers refer to control rod positions. Positions 1, 3, 5, 7 and 9 were in use at the time of the accident.*

## Operator training

One of the purposes of SL-1 was to provide training and experience to military personnel in reactor operations. While operation of SL-1 was the contractual responsibility of C-E, actual operating crews were military personnel. Operating personnel were selected from non-Commissioned ranks on the basis of "good" background and performance in mechanics or electronics courses conducted by the army. Training comprised an eight month course at Ft Belvior in Virginia, of which four months was devoted to the trainee's speciality and four months to reactor technology. This was followed by a 12 week field training programme at SL-1 of which about 60 per cent (288 hours) comprised "operator training". Written and oral examinations were administered by the army and, if successfully completed, were followed by oral and practical examinations administered by C-E. Formal certification was via memorandum from the C-E Operations Supervisor. Chief Operator qualification required at least six months experience as an SL-1 qualified operator, successful completion of written and oral examinations administered by the military and successful completion of an oral examination administered by C-E.

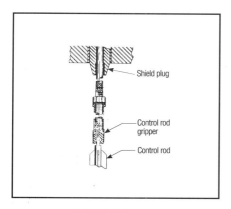

*Location of shield plug*

The military training programme was reviewed for content by C-E and deemed "adequate for the training of personnel for the operation of the SL-1". Operation of SL-1 by (typically) a two-man crew was on an around-the-clock basis. C-E supervision was limited to conventional working hours.

## Control rod drive disassembly

Access to the SL-1 core was via the nozzles for the control rods in the top plate. The control rod assembly comprised three pieces — the cruciform control rod blade with an upper cruciform extension, a control rod extension tube and the rack gear.

The cruciform extension of the control rod blade terminated in a boss, which was enaged by a gripper on the end of the

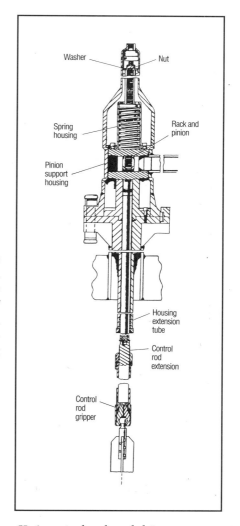

*SL-1 control rod and drive.*

control rod extension tube. The top of the rack gear terminated in a threaded stud. An integral shield plug/extension tube housing was fitted at each control rod position. Upon this is mounted the pinion support housing which, in turn, supported a spring housing. In normal operation a large washer and a castellated nut were screwed on to the threaded section of the rack gear. The washer bore against the top of the spring to limit the downward motion of the control rod.

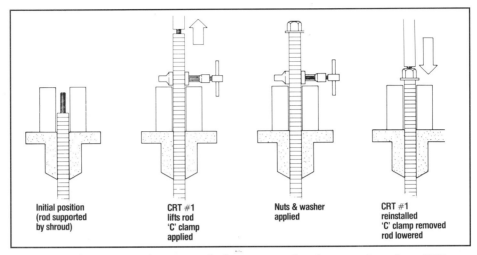

*Sequence of operations for removal of SL-1 control rod nut and washer. CRT was the acronym for a special lifting tool.*

Access to the SL-1 core via the control rod nozzles necessitated removal of the top spring housing, the pinion drive housing and the integral shield plug/extension tube housing. Before this could be done, of course, it was necessary to remove the nut and washer from the threaded section of the rack gear so that the spring housing and pinion support housing could be slid up and off the rack gear. This operation involved:

❑ Screwing a special lifting tool (essentially a metal bar with one end threaded internally to accept the threaded end of the rack gear) onto the threaded extension of the rack gear.
❑ Raising the whole control rod assembly approximately 10.16cm.
❑ Installing a "C" clamp on the rack gear to hold the control rod assembly in place.
❑ Removing the lifting tool and removing the nut and washer.
❑ Re-installing the lifting tool.
❑ Removing the "C" clamp and lowering the control rod assembly until the boss on the end of the control rod blade extension bore on the control rod shroud, supporting the weight of the control rod assembly.
❑ Removing the lifting tool.

In view of the critical nature of this operation, both figuratively and literally, it is interesting to note the actual written procedures upon which the operating crew had to depend to guide them in this work:

"9. Secure special tool CRT No 1 on top of rack and raise rod not more than 4 inches [10.16cm]. Secure C-clamp to rack at top of spring housing.

"10. Remove special tool CRT No 1 from rack and remove slotted nut and washer.

"11. Secure special tool CRT No 1 to top of rack and remove C-clamp, then lower control rod until the gripper knob located at upper end of element makes contact with the core shroud."

The reassembly instructions stated: "Assembly of the rod drive mechanism, replacement of concrete blocks and installations of motor and clutch assembly are the reverse of disassembly". These quotations are from the reactor operating instructions and appear to be the only written procedures available to the operators. There was no position stop on rod movement and it was possible for the operator to withdraw completely the 36.32kg rod assembly from the core. The 10.16cm withdrawal distance called for in the instructions was to be estimated by the operator.

An increase in the length of the threaded extension to the rack of about 10.16cm would have precluded the requirement for this kind of operation, since the action of unscrewing or screwing on the nut could have been used to transfer the load of the control rod assembly from the spring housing to the shroud and vice versa.

## Operations to 23 December 1960

C-E took over contractual responsibility for the reactor from ANL on 5 February 1959 to continue a programme to obtain operating experience, to develop plant performance characteristics, to obtain core burn-up data and to train military personnel in reactor operation and maintenance.

In reviewing the transfer of responsibility to C-E, the Army Reactors Branch of USAEC questioned the adequacy of the technical documentation provided by ANL and requested C-E to prepare revised manuals, as well as further to develop and modify such manuals in the light of experience. A "complete operating manual" prepared by C-E was approved by the AEC's Idaho Operations Office in March 1959.

During the initial period of SL-1 operations from January 1959 (when the reactor was still the responsibility of ANL) to April some control rod sticking was experienced — rods failed to drop cleanly into the core under gravity. This was tentatively ascribed to crud build-up on the rod drive seals, and ameliorative action was taken by ANL, which modified the seals, and C-E, which installed a filter in the seal water circuit.

The C-E programme commenced with test operations and cold critical experiments followed by a 1000-hour sustained power run, concluded in July 1959. When the reactor was shut down for inspection in August it was discovered that the boron strips were noticeably bowing out from the fuel assemblies. The AEC had requested C-E to evaluate the performance of the SL-1 core, and in the summer of 1959 C-E made the following recommendations:

❏ For "field application" use of aluminium as a core material was "not sufficiently advanced". Stainless steel was suggested as the preferred material.
❏ A new SL-1 core should be designed to have an adequate shut down margin with any one control rod removed.
❏ The control rod drive mechanisms should be redesigned.

These recommendations were accepted by the AEC and arrangements were made for the procurement of a new core, to be available by spring 1961. During an inspection by C-E in August 1960, extensive deterioration of the boron strips was noted. Pieces of boron strip were missing from some fuel assemblies and the assemblies in the centre of the core were extremely difficult to remove. Attempts to remove some assemblies resulted in pieces of the boron strips falling off. Numerous fragments of strip were subsequently collected from the bottom of the reactor vessel. Luke and Cahn of C-E reported at the time that in the case of one assembly "both the full boron strip and the half boron strip were missing" and went on to note that while it had been planned to remove five fuel assemblies for inspection, only three were removed "because the corrosion of the boron side plates was found to be so severe that further removal of elements could cause an unsafe condition in the reactor". Luke and Cahn calculated that about 18 per cent of the boron had been lost from the core and a shutdown margin of about 2% remained. They also calculated that loss of all boron would make the reactor supercritical by about 3.3% with all control rods fully inserted. Tests showed that withdrawal of the central control rod a distance of 36.32cm would be sufficient to establish criticality. The authors noted that: "The rate of loss of boron has been constant over the past 300MWd of operation. Indications are that this shutdown margin will continue to decrease, thus requiring remedial action".

The situation was reviewed by C-E Nuclear Division's Safety Committee in November 1960 and a recommendation by Luke and Cahn, to instal cadmium strips in two unused control rod channels, was adopted. The Committee also instructed that a detailed record of control rod positions should be maintained in order to reveal any future reactivity changes. The cadmium strips (with a total worth of about 1%) were installed on 11 November after which operation of the reactor at a higher power level than before (4.7MWt vs 3MWt) began.

From September 1960 onwards significant deterioration in control rod performance was experienced, this becoming particularly marked in November and December, when the SL-1 Operations Log recorded 33 cases of sluggish or sticking rods. In response the SL-1 Plant Superintendant (military) established full rod-travel exercises, to be carried out halfway through each operating shift. This order was temporarily rescinded on 22 December to accommodate a special power run to obtain equilibrium data. On 23 December SL-1 was shut down for maintenance. It was planned to bring the reactor back to power on 4 January 1961. During the shutdown period 44 cobalt flux measuring wires were installed between the plates of the fuel assemblies, a procedure necessitating the removal of the control rod shield plugs.

On 3 January 1961 the day shift (08:00 - 16:00) disassembled the control rod drive mechanisms, removed the shield plugs and installed the flux measuring wires. The next shift (16:00 – 00:00) was assigned to reassemble the control rod drives and prepare the reactor for restart. The operating crew comprised three people, a qualified Chief Operator, a qualified Operator-Mechanic and a trainee, all military personnel.

## Accident, 3 January 1961, and subsequent investigations

At 21:01 on 3 January an automatic heat detection alarm in the SL-1 building began transmitting a continuous signal to three locations at the National Reactor Testing Station (two fire stations and the Communications Building). At about the same time the personnel monitor in the Gas Cooled Reactor Experiment gatehouse (1.448km north-east of SL-1) alarmed with all meter needles on full scale deflection, and could not be reset.

A fire crew was sent to SL-1 and by 21:15 the Assistant Fire Chief had investigated the Support Facilities Building as far as the foot of the stairway leading up to the reactor operating level, where radiation levels of about 250mRem/h were measured. During subsequent building entries fields of 200Rem/h at the first landing on the stairway to the operating level and over 500Rem/h at the doorway to the reactor operating level were measured.

At 21:45 two C-E personnel made their way to the entrance to the operating level and saw two of the operating crew on the floor, one moving. At 22:50 a five-man team entered the operating level and removed the crewman who was still alive. This man was subsequently pronounced dead at 23:14. At 23:00 a four-man team entered the control level and located the third member of the operating crew lodged in the ceiling above the reactor vessel in a field subsequently calculated to be about 1000Rem/h.

By 20:15 on 4 January the second victim had been removed, and the third by the evening of 8 January. Subsequent operations included:

❏ Confirming and maintaining subcriticality.
❏ Photography of the operating room floor area and reactor internals.
❏ Decontamination.

❏ Removal of reactor vessel to hot cells for examination.
❏ Dismantling of structure.
❏ Complete decommissioning of the site and burial of all contaminated material.

Photographic and later laboratory investigation revealed that 20 per cent of the SL-1 core was destroyed together with 47 per cent of the fuel in the central 16 fuel assemblies. The core was extensively damaged, with recognizable fuel assemblies severely distorted and displaced radially outwards. The four outer control rods were locked in their shrouds, fully down. The shrouds were twisted, flattened and displaced outwards. The central control rod blade, with its shroud collapsed against its lower end, was lying horizontally across the top of the core. The rod was firmly bound in the shroud and had not been moved after the shroud had collapsed, as evidenced by the imprint of the shroud pressure equalization holes on the rod blade and spattered molten metal on those areas only.

The central rod, and only the central rod, was withdrawn from the core (by about 50.80cm) at the time of the transient. All piping connections to the reactor vessel were sheared at the vessel outer face, and the vessel was bulged from an original circumference of 4.31m to 4.41m in a region corresponding to the top of the core. There was a more severe bulge at the top of the vessel, where the circumference had increased to 4.63m. All the loose shield plugs (Nos 1, 3, 4, 7 and 9) had been ejected and, with the exception of the No 9 plug, had lodged in the ceiling.

The No 7 plug had impaled the operating crew member who was found lodged in the ceiling.

The No 9 plug, with the pinion housing and spring housing attached, was found lying across the reactor's top plate. Deformation marks on the flange of this plug matched other marks in the ceiling, indicating that this plug had entered the ceiling before falling back down on to the reactor. The extension tube housing was collapsed about the control rod extension. A short (27.94cm) section of the rack gear was still attached to the control rod extension. The remaining section of the rack gear, with nut and washer attached but with the threaded extension broken at its cotter pin hole, was found on the fan room floor. The broken end of this stud was found still in place in the control rod lifting tool. A broken "C" clamp, with its opening set to the rack diameter, was also found on the fan room floor. From this it is clear that at the time of the accident the central (No 9) control rod shield plug, pinion housing and spring housing had been reassembled, the control rod raised and clamped, the nut and washer put back on and the handling tool attached. All that remained was to remove the clamp and lower the rod to its usual position.

The cause of the SL-1 accident was the manual withdrawal of the central control rod a distance of about 50.80cm. Calculations suggest that a withdrawal of the central control rod by 42.42cm would establish criticality. Continuing the withdrawal to 50.80cm would have added about 24mk. Calculations suggest that the reactor reached a power of about 20 000MWt, at which point a proportion of the fuel in the centre of the core (about 5 per cent) vapourized causing steam production and violent destruction of the central core region. The formation of a steam void terminated the nuclear transient but created a high pressure region. The pressure wave front first struck the vessel wall next to the core (causing the bulging), then the bottom of the vessel, then finally accelerated upwards the mass of water above the core. The water level in the vessel was about 0.76m below the lid, so the water slug had this distance over which to acquire kinetic energy. The slug hit the vessel lid with sufficient force to collapse the extension tube housings around the control rod extensions (estimated to require about 10 000lb/in$^2$) and to bulge the pressure vessel. The loose shield plugs were ejected at an estimated velocity of 93.3km/h and penetrated the ceiling.

They were followed shortly after by the reactor vessel as the transfer of energy from the water to the lid caused the vessel to rise 2.74m out of its housing and hit the ceiling. The vessel then fell back to its original position.

The entire sequence, from the start of control rod withdrawal, took between two and four seconds. Evidence for the movement of the reactor vessel is quite conclusive. Insulating blocks used well down the vessel sides were found on the operating level floor, and the drive shaft coupling on the No 5 control rod mechanism (which remained bolted to the reactor lid) showed evidence of collision with the drive shaft of the overhead crane above the reactor. In addition, samples taken from the operators indicate that they were exposed to an integrated thermal neutron flux of the order of 10$^{10}$ n/cm$^2$ and an integrated fast neutron flux as high as 10$^{13}$ n/cm$^2$, which is only explicable if "somehow an essentially bare core were brought into the vicinity of the victims fast enough that delayed neutrons were still being emitted in copious quantities" [Thompson and Beckerley].

The reason for the withdrawal of the control rod by the operator is unknown. Suggestions ranging from attempts by the operator to free a stuck rod to mental instability have been made, but they all remain speculative.

## Discussion

The SL-1 accident is of interest less because of the actual initiating event than because in the operating life of the reactor a number of deficiencies became apparent, were recognised, yet did not prompt action to halt SL-1 operation pending any kind of review. Forwarding his Final Report to the USAEC, General Manager Nelson notes in his letter of transmittal:

"It is known that certain undesirable conditions had developed with respect to the reactor and its operation, some having their origin in the design of the reactor and others in the cumulative effects of reactor operation, which do not now appear to have had a direct relation to the immediate cause of the incident. The Board observes, however, that the over-all effect of these conditions produced an environment in which the possibility of an incident may have increased beyond that necessary".

There were early concerns expressed about the quality of documentation, with C-E requested to prepare revised manuals and to develop and modify them in light of operational experience. Nelson, as Chairman of the General Manager's Board of Investigation of the SL-1 incident, noted in his 10 May 1961 Report to the Joint Committee that "reactor operating procedures completely satisfactory to the USAEC have never been completed by Combustion Engineering, although they have been in the process of preparation and revision since mid-1959". However, the Operating Manual, which included the very brief instructions for control rod drive disassembly, was approved by the AEC's Idaho Operations Office in March 1959.

In summer 1959 C-E informed the AEC, in apparently quite unambiguous terms, about the unsuitability of the core structural materials, the unsatisfactory nature of the control rod configuration and the problems with the control rod drive design. The company's views were endorsed by the AEC and arrangements were made to obtain a redesigned core. It seems reasonable to argue that at that moment future operations of SL-1 should have been rigorously reviewed, particularly with respect to its functions in the areas of "demonstration and training".

At the very least the question of 24-hour supervisory support for the operating crews should surely have been reconsidered. The SL-1 project manager pointed out at the hearings on the accident that "the contractor's operating budget for the SL-1 operation did not permit staffing of the plant on an around the clock basis", and added that C-E's proposal for increased staffing levels to support an expanded testing programme had not been approved by the AEC for budgetary reasons. The point could be made that SL-1 was simply the first version of what was to be a standard military machine, so unsupervised operation by military crews was a necessary and appropriate part of a realistic training programme, but this point would lose much of its force after the decision had been made essentially to redesign the reactor. The problems posed by the control rod configuration had been identified and the possibility of future core deterioration had been raised. These factors, combined with the extent of training received by the operators and the nature of the documentation upon which they had to rely, suggest that professional supervisory support on a 24 hour basis would have been a prudent step. By the summer of 1960 significant core deterioration in SL-1 had been discovered. The subject was reported in the literature, reviewed by a C-E safety committee and, indeed, was reported in the November issue of the nuclear industry periodical *Nucleonics*. Not only was the response to this limited to the installation of cadmium strips, but operation of of the reactor was recommened at a higher power level than before without an adequate hazards review.

Nelson stated at the post-accident hearings that C-E "routinely and consistently forwarded reports on reactor operations" to the USAEC Idaho Office's Military Reactors Division and that the Director of that Division or his SL-1 Project Engineer made frequent visits to the facility. Some AEC staff should, therefore, have been aware of the deteriorating condition of the SL-1 core.

The question of the sticking control rods is somewhat different. Not only did Nelson note that: "The sticking of the control rods ... was not considered a malfunction and therefore not specifically reported", but the SL-1 Project Manager stated that he was unaware of this problem, and if he had known he would have considered it grounds for shutdown, pending resolution of the situation. In addition, Nelson, in discussing periodical inspections of the SL-1 facility by AEC staff, stated that these appraisals of the safety of the plant "although comprehensive in most respects, did not include inspection of the nuclear safety of reactor operations". On the other hand, nowhere in the contemporary literature, or in this study, is there any suggestion that C-E or any of the USAEC Directorates were guilty of deliberately concealing information, or acting in any dishonest or unethical manner in order to maintain in operation a reactor they knew to be unsafe. All individuals involved in the SL-1 project appear to have been working with enthusiasm to complete a project, with which initial experience had been good, within challenging time and budget constraints.

Thompson makes a most perceptive observation when he comments:

"It is clear, and many people have later said so, that the reactor should have been shut down pending resolution of the boron difficulties and the general deterioration of control rod operation. In fact no one did so or even brought the malfunctions to the attention of any responsible safety group. In the climate that existed before the accident, it is likely that if one man had decided that the reactor should be shut down for safety reasons, he would have been ridiculed and would almost certainly have had an unfriendly response since he would have had to say some rather harsh things to accomplish his purpose". [Thompson and Beckerley]

A root problem with the SL-1 project was the fact that the large numbers of organisations involved and the extreme complexity of the lines of responsibility resulted in a total failure to establish clear and continuous responsibility for nuclear safety. Design and commissioning of SL-1 was carried out by Argonne National Laboratories, at which point responsibility for co-ordination and direction was held by the Programmes Division of the USAEC Chicago Operations office. Subsequent operation by C-E came under the aegis of the USAEC Idaho Operations office, with the USAEC Division of Reactor Development and its Army Reactors Office exercising overall programme responsibility. The Idaho Operations Office and its Military Reactors Division, as well as the Army Reactors Office in Washington, participated in decisions about the amount of supervision to be used at the reactor. Yet, while there were numerous people with some safety responsibility at various periods in the reactor's life, there was no single group with continuous safety responsibility, awareness of the growing problems and the authority to take decisive action.

Thompson identifies four basic organizational principles of reactor safety:

- As far as possible design, construction and operation should be the responsibility of a single organization.
- Responsibility for safety and all facets of reactor operation should be unequivocally defined ("a line organization should be used, not a committee").
- Safety review should be carried out by a (single) competent group external to the operating organization.
- Reviews repeated by competing safety groups can "unduly harass the operating group and thereby reduce safety".
- The ultimate responsibility for operational safety must ultimately rest on the immediate operating team at the reactor.
- "In the final analysis the reactor shift supervisor and, in turn, the operator at the control console should have the authority to shut down the reactor if either believes it to be unsafe". [Thompson and Beckerley].

More than a quarter of a century after the event these principles may not be new, but they bear repeating.

## Bibliography

US Atomic Energy Comission, SL-1 Report Task Force, *IDO Report on the Nuclear Incident at the SL-1 Reactor*, IDO-19302, January 1962.

Hearings before the Joint Committee on Atomic Energy, Eighty-Seventh Congress, First Session on Radiation Safety and Regulation, June 12, 13, 14 and 15, 1961, US Government Printing Office, Washington, 1961.

T J Thompson, and J G Beckerley, (eds), *The Technology of Nuclear Reactor Safety*, MIT Press, 1964, Chapter 11, "Accidents and Destructive Tests".

C W Luke, and H Cahn, *Evaluation of the Loss of Boron in the SL-1 Core*, C-END-1005, Combustion Engineering Inc, 1960.

SL-1 Recovery Operations, C-END-1007 (IDO-19301), Combustion Engineering, 1961.

*Final Report of the SL-1 Recovery Operation*, May 1961 through July 1962, IDO-19311, General Electric Company, 1962.

C A Nelson, et al, *Final Report of the SL-1 Board of Investigation*, United States Atomic Energy Commission, September 1962.

D Mosey, and K Weaver, *A Discussion of Institutional Failure and its Importance in Nuclear Safety*, Ninth Annual Conference of the Canadian Nuclear Society, June 13–15, 1988.

# FERMI 1

The Fermi 1 plant was built in the late 1950s. Located near Munroe, Michigan, the plant was owned and operated by the Power Reactor Development Company (PRDC), an organization of 21 utilities and manufacturers. Research and development and subsequent design was carried out by Atomic Power Development Associates, an organization of 42 utility, manufacturing and engineering interests.

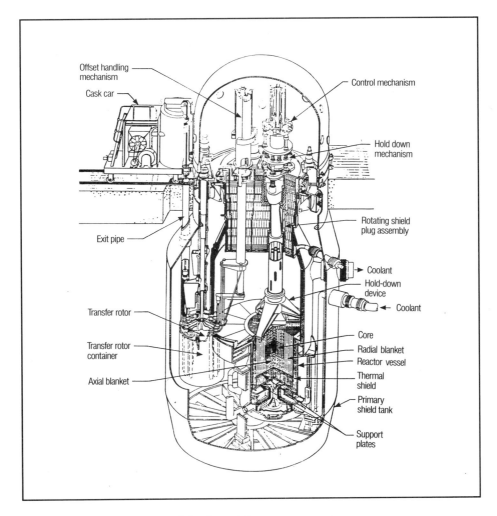

*The general arrangement of the Fermi 1 reactor.*

The reactor was a 300MWt liquid metal cooled fast-breeder. Heat was removed from the reactor core and breeder blanket by the liquid sodium primary coolant, and transferred first to the secondary liquid sodium heat transport circuit in three parallel intermediate heat exchangers, and finally to a steam/water circuit via three once-through steam generators.

The primary sodium coolant flow rate was $13.2 \times 10^6$ lb/h with a design inlet temperature of 260.0°C and an outlet temperature of 426.7°C. Outlet pressure was slightly above atmospheric. Three discrete, parallel primary and secondary coolant loops were provided.

The core was in the form of a vertical cylinder 83.1cm in diameter by 76.2cm high made up of 91 fuel assemblies surrounded by a 237.7cm diameter by 199.4cm high cylindrical array of depleted uranium fuel assemblies forming the radial and axial breeder blankets.

Each core fuel assembly was a 6.6cm square, 12 x 12 array of 140 fuel elements plus four stainless steel tie rods surrounded by a stainless steel can. The fuel comprised uranium/10 wt per cent molybdenum enriched in U-235 to 25.6 per cent. The cladding was 0.375cm diameter zirconium tubing. About 25 of the core positions were provided with thermocouples.

Sodium flow from the three coolant loops passed through an inlet plenum, over a conical flow guide and upwards through the core.

The lower portion of the inlet plenum and the conical flow guide were covered with six 0.102cm thick segments of zirconium sheet to provide an added safeguard against penetration of the reactor vessel by molten fuel in the event of a loss of coolant accident. This design feature was added at a comparatively late stage in construction. Reactor control was via two boron carbide rods, one for regulation and the other for shim. Automatic control throughout the power range was available. The shutdown system used eight spring driven boron carbide shutdown rods, with an insertion time of less than one second.

## Chronology of a fuel melt

Construction of the power plant was completed in 1963 and first criticality was achieved in August of that year. Initial operation was at power levels up to the maximum authorized level of 1MWt. In December 1965 authorization was given for operation at power levels up to 200MWt (66MWe). By July 1966 power had been raised to 100MWt and in early August a 60 hour test run at this power level was completed. In June 1966, during tests at 67MWt, abnormally high temperatures (20 to 25 per cent above normal) were detected in two fuel assemblies.

The temperatures returned to normal during the initial period of 100MWt operation in July, but during a subsequent 60 hour test run in August they went up again, this time to between 40 and 47 per cent above normal for the pertaining power level and coolant flow rate. Nevertheless, the temperature readings were lower than those for the central core fuel assemblies and all temperatures were below the range anticipated for full power operating conditions. Also, in addition to the two high temperature fuel assemblies, one assembly exhibited anomalously low temperatures.

Operation at 65MWt in September confirmed these anomalous temperature readings, so the three subassemblies were moved to locations which had previously exhibited normal temperatures to establish whether the problem lay in the fuel assemblies themselves or in the thermocouples.

Reactor operations resumed on 5 October with the objective of obtaining more test information at the 67MWt level, including information on fuel assembly outlet temperatures with the rearranged assemblies. Power was brought up to the 8MWt level by 14:20 and the reactor was then placed on automatic control. At 15:00, with power at 20MWt, a control signal indicating an erratic rate of change of the neutron level was noticed but was attributed to noise pick up in the control system. The reactor was switched over to manual control, after which the instability disappeared.

Power increase under automatic control was resumed and at about 15:05, with the power at 27MWt, the erratic control signal reappeared. Shortly afterwards it was noticed that both regulating and shim rods were withdrawn further than normal — both rods were withdrawn about 22.9cm compared with a normal operating position subsequently established to be 15.2cm.

A check of the coolant exit temperatures revealed that two fuel assemblies had significantly higher than normal outlet temperatures, but the power increase under automatic control was continued.

The radiation monitors in the containment building ventilation exhaust ducts alarmed at 15:09 and the building was automatically isolated. The area radiation monitor in the fission product detector building also alarmed and isolated that building. The reactor power increase was halted at this time with reactor power at 31MWt, and power reduction began. At 15:20 reactor power had reached 26MWt and the reactor was manually tripped.

## Post incident investigations

Post incident investigations were protracted. Great care had to be taken not only to avoid any action which might bring about secondary criticality, but also to minimise further damage to the core and consequent destruction of valuable evidence.

Analytical and experimental work performed immediately following the incident determined that a significant reactivity loss had occurred. Approximately two thirds of this was attributed to fuel relocation and one third to distortion due to overheating. Using the time history of the shim and regulating rods, it was determined that an anomalous negative reactivity insertion, which had been automatically compensated for by control rod withdrawal, began when the reactor reached 5.5MWt and continued at a steady rate until power reached 18MWt, at which level the rate of negative reactivity insertion jumped sharply and continued at an accelerated rate until 30MWt was reached. This suggested that a flow blockage occurred before the rise in power, and that thermal distortion of the core began at 5.5MWt and fuel melting at 18MWt.

Between January and May 1967 four fuel assemblies were removed from the reactor for laboratory examination but they failed to reveal sufficient information either on the cause of the incident or the extent of fuel damage. Two fuel assemblies (M-217 and M-098), which were known from visual inspection to be stuck together, were split apart using remote tooling and removed from the core in early August 1967. Subsequent examination revealed that extensive melting and relocation had taken place, to the extent that the centres of gravity of the two assemblies had moved downwards by 1.90cm and 4.76cm. Further inspections revealed that an adjacent assembly was badly bowed from overheating. A fourth assembly showed no external damage save for swollen end caps on seven fuel elements.

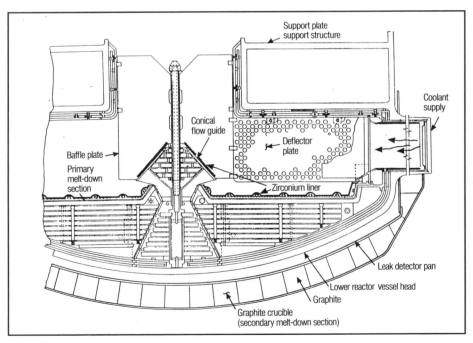

*The lower portion of the Fermi 1 reactor showing the zirconium liner plates. The plates on the flow guide were added late and not recorded on the "as-built" drawings.*

Thermal-hydraulic analyses indicated that the two badly damaged assemblies experienced coolant flows reduced to between 1 and 3 per cent of nominal while the second two experienced flows of 7 and 30 per cent respectively.

## Establishing the cause

The cause of the flow blockage was not immediately apparent and a large number of possibilities could be postulated ranging from the presence of a gas bubble to cladding fragments blocking flow at the support grids.

In September 1967, following draining of the reactor vessel, a segment of what appeared to be zirconium liner from the conical flow guide was discovered on the bottom of the inlet plenum. By March 1968, this had been retrieved and positively identified. Preparations were made to remove the remaining segments, which were assumed to be still attached to the flow guide, and in November of 1968 it was discovered that a second zirconium segment was missing. By the end of December this missing segment had been located lodged against the underside of the lower core support grid, and removed. Subsequent hydraulic tests confirmed that the coolant flow blockage had indeed been caused by one of the loose zirconium plates.

Each plate was held in place by three zirconium machine screws which were then tack welded to the plates. Flow induced vibration, it seems, eventually caused two plates to break free of their mounting screws. One of the plates was carried upwards by the coolant flow to the underside of the fuel asembly support structure where it sufficiently restricted flow to cause fuel melting.

A most important feature of the Fermi incident was that it focussed attention on the importance of preventing channel/subchannel flow blockage. It also showed that under severely reduced flow conditions, while fuel might melt, the molten fuel re-solidified after moving a relatively short distance.

The two fuel assemblies which actually melted experienced a very considerable flow reduction (1 to 3 per cent of nominal) while the assembly that distorted but did not melt experienced 7 per cent of nominal flow. Damage to the fourth assembly (30 per cent of nominal flow) was trivial.

## Discussion

At the time of the incident the operators took no action to shut down the reactor until the building ventilation exhaust monitors alarmed at 15:09, nine minutes after the first erratic neutron rate signal was noted, and the reactor was not actually tripped until 15:20. In fact, the operating crew did not have positive guidance on whether or not to shutdown in the event of anomalous reactivity behaviour or fuel assembly outlet temperature deviations. For subsequent reactor operations this situation was improved through both technical means (provision of an on-line computer to provide prompt information on control rod position) and procedural means (establishment of appropriate operator action criteria).

Installation of the zirconium plates took place comparatively late in the construction phase of the reactor in response to concerns expressed by the Advisory Committee on Reactor Safeguards. The late installation, combined with inadequate drawing revision procedures, resulted in the "as built" drawings giving no indication of the existence of the liner plates — a fact which significantly delayed positive identification of the loose plates when they were located.

Failure to document properly design changes or modifications in the course of construction represents a notable failure in nuclear safety management with potentially serious consequences. The later in the construction process the modification is made, the greater the risk of an inadequate analysis of its possible safety impacts, while the installation itself may not be subjected to sufficiently rigorous quality assurance scrutiny.

A perhaps less definable institutional problem can be discerned from the rationale for installation of the liner plates in the first place. It appeared that PRDC was faced with the choice of spending a modest amount of money to provide the additional engineered safeguard or making quite considerable efforts to justify to the ACRS why it should not do so.

As it turned out, subsequent analysis showed that the liner plates were unnecessary and they were not replaced when reactor operations resumed. If the principle that the owner/operator of a nuclear power plant is responsible for the safe and reliable operation of that plant is applied to this situation, then it could be argued that PRDC did not fully discharge their responsibility for safe design, construction and operation. They failed to carry out sufficient analysis of a proposed modification to assure themselves, and the regulatory authority, either that the modification was desirable and would not adversely affect safe and reliable plant operation, or that it would confer no palpable safety benefit and should not be implemented.

## Bibliography

Atomic Power Development Associates, *Enrico Fermi Atomic Power Plant Description*, USAEC Report APDA-124, January 1959

R L Scott, *Fuel Melting Incident at the Fermi Reactor on October 5, 1966,* Nuclear Safety, 12–2, March—April 1971

J G Duffy et al, *Investigation of the Fuel Melting Incident at the Enrico Fermi Atomic Power Plant*, Proceedings of the American Nuclear Society National Topical Meeting on Fast Reactors, San Francisco, April 10–12, 1967, Report ANS-101, pp 2–15 — 2–37

Atomic Power Development Associates, *Report on the Fuel Melting Incident in the Enrico Fermi Atomic Power Plant on October 5 1966,* USAEC Report APDA-233, December 15, 1969

A J Friedland, *Thermal-Hydraulic Analysis of the October 5 Fuel Melting Incident in the Enrico Fermi Reactor,* USAEC Report APDA-LA-5, Atomic Power Development Associates, July 1969

Transcripts from JCAE Hearings, January and February, 1968

# LUCENS

The Lucens reactor was an 8MWe (28MWt) heavy-water moderated, carbon dioxide cooled experimental power reactor located underground.

The Lucens facility was owned by SNA (Société Nationale pour l'Encourage-ment de la Technique Atomique Industrielle), constructed by ThA (Therm-Atom) and operated by EOS (Western Switzerland Power Company). Site preparation began in the summer of 1961, rock excavation was conducted in 1962 and 1963, erection of the electro-mechanical equipment started in 1964 and the station achieved first criticality in December 1966. The reactor was severely damaged on 21 January 1969 as a result of an in-core loss of coolant accident (LOCA) caused by subchannel flow blockage and consequent fuel melting. The following account

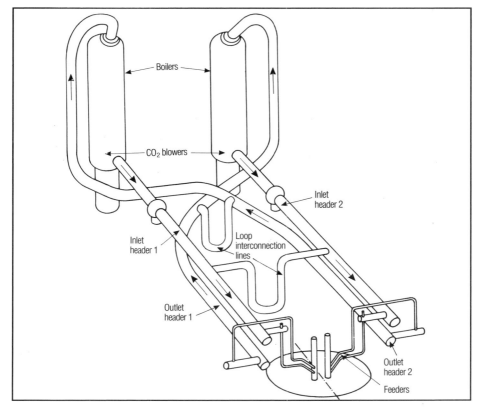

*The Lucens primary circuit.*

is based principally on the Swiss Commission of Enquiry's Report on the accident and correspondence with EOS.

## Plant layout

The Lucens plant was mainly housed in three adjacent underground vaults — the reactor vault, the machine vault and the fuel storage vault. The reactor vault contained the reactor itself, the primary heat transport circuit, the reactor auxiliary systems and the fuelling machine. The machine vault contained the turbine-generator and its associated equipment, together with the other conventional (ie non-nuclear) plant equipment. The fuel storage vault contained the irradiated fuel storage bay and the fuel handling control room. The fuel storage vault and the reactor vault were linked by a fuel transfer duct, and a connecting tunnel linked

*Cutaway view of the Lucens reactor.*

the reactor vault to the machine vault, which was isolated from the nuclear portion of the installation by airlocks. A tunnel from the machine room ran to the surface where the service building, ventilation plant and main control room were located.

## Reactor description

The reactor comprised a vertical cylinder about 3m in diameter by 3m high (the calandria) filled with heavy water and threaded by 87 vertical calandria tubes containing the re-entrant fuel channels and control and shut-off rods.

The reactor was cooled by $850lb/in^2$ carbon dioxide circulated through a "figure of eight" coolant loop comprising two steam generators, two blowers, and two inlet and outlet headers. From inlet header No 1 the gas was distributed by

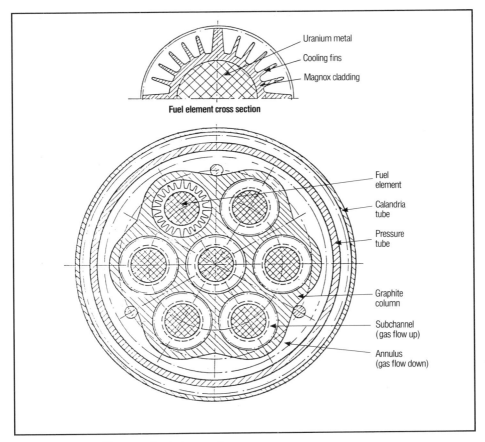

*Cross-section of Lucens fuel channel.*

feeder pipes to half the core. The hot gas from the fuel channels flowed through outlet feeders to outlet header No 1 and to the top of steam generator No 2. The cooled gas then flowed through blowers to inlet header No 2, the other half of the core, outlet header No 2, steam generator No 1 blowers and inlet header No 1. Gas balance lines linked each pair of inlet and outlet headers.

Gross coolant flow through the core was automatically controlled by butterfly valves in the inlet headers to match flow to reactor power and to maintain a constant outlet temperature. In addition, flow to individual channels could be controlled through valves at the inlet feeder/header junctions. Individual channel outlet temperatures and flow rates were monitored at the outlet feeder/header junctions.

Each fuel assembly, contained in a 125mm OD zircaloy pressure tube, comprised a graphite column made up of a stack of three graphite blocks, pierced by seven vertical 32mm diameter subchannels. For the purposes of this discussion the term "fuel assembly" refers to the whole assembly of graphite blocks and 28 fuel rods, and "fuel element" to an individual rod. Fitted into each subchannel was a stack of four 17mm diameter uranium metal fuel elements contained in finned magnesium alloy tubes. Three zircaloy tie rods and an arrangement of springs held the whole assembly together.

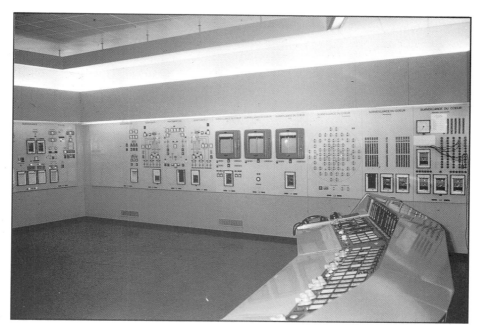

*Control room, showing the core monitoring panels and blower auxiliaries.*

The pressure tubes were closed at their lower ends so that gas flow was down the annulus between the graphite stack and the pressure tube wall and up the seven subchannels. The fuel was enriched to 0.96 per cent U-235.

Refuelling was accomplished off-line, using a fuel handling machine located immediately below the reactor. Fuel assemblies were removed complete with their surrounding pressure tubes.

Control was via four control rods and six shut-off rods, the calandria tubes for the latter being designed specifically to resist calandria overpressures resulting from an in-core loss of coolant accident (LOCA).

## Operating experience 1966–68

The trained operating crew was on site for much of 1966. This helped in electrical construction work and in the commissioning of the plant. The crew prepared all the operating procedures together with the details of the experimental programme.

First criticality was achieved in late December with a reduced fuel loading comprising only the 36 fuel elements in the central zone of the core. After checking the physical characteristics of this reduced loading the reactor was defuelled. Hot commissioning resumed at the beginning of 1967. Almost a year was allocated for commissioning work following which, in late 1967, station trials at progressively increasing reactor power levels began.

Following a ten day uninterrupted run at high power (21MWt and above), in May 1968 EOS took over operating responsibility. An extensive testing programme was conducted, including loss of feedwater with automatic emergency cooling, and full station blackout with cooling by natural circulation in the primary, secondary and emergency cooling circuits and decay heat removal via evaporation of a water tank on the roof inside the reactor vault.

Between mid-August and the end of October 1968 the reactor operated continuously at thermal power levels between 22 and 28.5MWt, including a short period at the full rated power level of 30MWt.

At the end of October the unit was shut down for maintenance. A major maintenance concern related to primary circuit leak-tightness and, as well as replacing or repairing a number of valves, considerable attention was devoted to the shaft seals on the main circulating blowers. These seals comprised a labyrinth gland, a water lubricated graphite-metal rotating seal and several floating rings. The rotating seal used the "water-film blocking" principle whereby a continuous flow of water was maintained between the metal and graphite surfaces. To prevent contamination of the primary circuit with moisture, a flow of $CO_2$ ("blocking gas") from the primary circuit was maintained through the labyrinth gland to entrain the water.

On a number of occasions, however, water from the rotating seal had entered the primary circuit through the labyrinth gland. Remedial work included replacement of the rotating seal (after mechanical failure of the graphite portion) and modifications to the gland and seal systems. In spite of this work, moisture contamination of the primary circuit reappeared sporadically during the August-October period of operation.

In the course of the maintenance shutdown repeated seal modifications and subsequent tests gave indeterminate results. By early 1969 it had been concluded that the lubricating grooves in the metal seal face allowed an unacceptable flow of water and the most practical approach was to accept a possible reduction in service life and reverse the metal seal element so that it presented a smooth face to the graphite. This appeared to bring a dramatic improvement — indeed, in the case of one seal no measurable water leakage was observed.

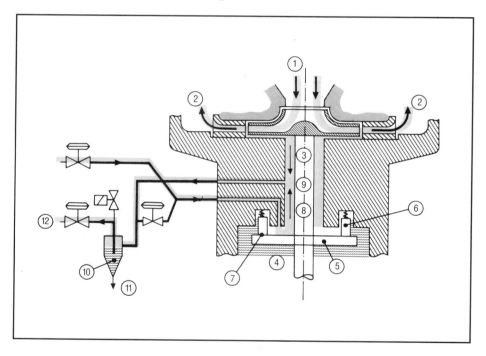

*Simplified drawing of blower seal with flowpaths.*

*1 — primary gas enters blower. 2 — compressed carbon dioxide exits blower wheel. 3 — carbon dioxide flow through upper labyrinth. 4 — demineralised seal-water at carbon dioxide pressure. 5 — rotating copper disc, nickel plated, with a chromium layer. 6 — springloaded graphite ring. 7 — lubricating and cooling water film between ring and disc. 8 — clean carbon dioxide purging lower labyrinth (not for all operating conditions). 9 — mixture of seal water and purge carbon dioxide. 10 — seal water separator. 11 — seal water to liquid waste treatment plant. 12 — purge carbon dioxide to dryer plant.*

In autumn 1968 the operator applied for a full operating licence. The safety authorities, in considering the request, stressed the weakness of the fuel assemblies, but had confidence in the containment measures taken and consented to deliver the operating permit.

A permanent operating licence for the station was granted to EOS on 23 December 1968 and preparations were made to resume continuous steady power operation for a period of about 200 equivalent full power days (EFPD), this being the remaining burn up limit of the fuel charge.

The only remaining question was whether the metallurgical limits of the fuel would be reached, with swelling the major concern.

## Accident: 21 January 1969

Preparations for restart began at Lucens on 6 January, following completion of work on the blower seals. After the blowers were returned to service, moisture levels in the main gas cooling circuit were found to be very high. The system was dried out by repeated flushing with dry compressed air and, on 18 January, the primary circuit was filled with dry $CO_2$.

The reactor was made critical at about 03:00 on 21 January and by 05:43 had been brought to a power level of 5MWt. At this point the burst cladding detection system gave indications of significantly higher than normal radioactivity readings in the coolant gas at three locations. The individual channel outlet flows and temperatures were checked but appeared to be within the normal range.

Reactor power was increased to 7MWt (10:30) and then to 9MWt (11:42) and individual channel flows and temperatures were again checked and again found to be within the normal range. However, some concern was felt about the radioactivity levels indicated by the burst clad detection system, although the notably high readings mentioned above appeared only momentarily.

At 17:03 another power increase to 12MWt was begun, this level being reached at 17:15. At 17:20 the reactor tripped on high coolant activity. Position indicators in the control room showed that all ten rods (shut off and control) had fully inserted and that the vault ventilation ducts had closed automatically.

Almost simultaneously, a large number of other alarm indications were received in the control room including low coolant pressure, low coolant flow, high vault activity, high calandria tank pressure, low moderator level, high fuelling machine sump level and high stack beta and gamma activity. From these and other indications the operators deduced that there had been a large in-core failure in the primary circuit, causing one or more of the calandria rupture discs to relieve and ejecting heavy water into the vault.

Over the next 10-20 minutes the operators confirmed that fuel temperatures were falling rapidly (despite the shock wave the temperature gauges on the five instrumented fuel assemblies remained operational) and that the primary circuit system pressure had stabilized (at about 18lb/in$^2$). A radiation survey team went to the machine vault and confirmed high fields in the vicinity of the equipment lock accessing the reactor vault.

The emergency ventilation system, which was fitted with charcoal filters, was turned on at 17:58, then switched off since the stack iodine monitor had gone offscale and there was concern that environmental release limits might be violated. Since air beta-activity levels in the control room were increasing to above the maximum permissible concentration for occupational exposure, at 18:15 personnel moved to an adjacent building leaving a skeleton crew of two or three equipped with airmasks, an arrangement which continued until noon the following day when the control buildings had cleared.

Environmental monitoring conducted between 20:30 and 21:30 on 21 January confirmed that such emissions as had taken place were well below the regulatory limits (in some cases little above natural background) and principally comprised short-lived isotopes: Kr-88 and Rb-88.

## Recovery and accident investigation

The principal immediate recovery work involved regaining access to the vaults. Starting at about 16:30 on 22 January the emergency filtered air system was operated and the vaults flushed through with fresh air. Environmental monitoring teams downwind of the stack detected no significant releases as a result of this operation. By midday the following day (23 January), the machine vault and the fuel storage vault had been cleared and normal (ie unfiltered) ventilation of these areas could be resumed. The access tunnel and the machine room were now routinely accessible.

The extent of the spread of radioactive contamination through the vault complex indicated a significant leak from the reactor vault. During the night of 23–24 January, inspection revealed a leak of several cubic metres per hour at a cable penetration in the vicinity of the equipment airlock giving access to the reactor vault. This leak was stopped.

At the time of the accident the reactor vault became pressurized to about 3.2lb/in$^2$ following the release of the coolant gas. Over the following three days this pressure rose somewhat, principally due to instrument air in-leakage. On 25 January the reactor vault was first depressurized, then air flushed via the emergency filtered air system. Again, environmental monitoring detected no significant radioactive release. Following this, about 5000 litres of spilled D$_2$O were pumped from the fuelling machine vault sump to a holding tank.

An exhaustive investigation was undertaken and lasted some years. Central to this investigation was the removal and laboratory examination of a number of pressure tubes and fuel assemblies. The pressure tube of fuel assembly No 59 was severely damaged. Starting at about 60cm from the top of the pressure tube and facing subchannel No 1 was a 58cm long axial tear. About 12cm below this was a 4cm long bump with a maximum height of 4mm. About 60cm from the bottom of the tube and facing subchannel No 4 was a 15cm long by 3cm wide oval hole with melted lips. At this point too the pressure tube had undergone local circumferential swelling of about 5cm. Some 25cm above this, and in line with it, was a 35mm diameter semi-circular hole with melted lips. Close to the bottom of the tube were two brittle cracks 70mm and 110mm long facing subchannels Nos 6 and 1 respectively. From below the axial tear the inner wall of the pressure tube was oxidised and coated with a thick deposit of foreign matter containing pieces of graphite, a wafer of uranium and a fragment of zircaloy tie rod. The bottom third of the pressure tube contained graphite fragments, a coarsely granulated mass of brownish black fusion and combustion products and a solidified mass of uranium, subsequently determined to amount to about 16 per cent of the original uranium inventory (28 elements totalling 77.6kgU) of the fuel assembly. Fuel element fragments recovered from the calandria showed signs of combustion of the metallic uranium and of the magnesium cladding.

Fuel assembly No 49, adjacent to No 59, was relatively intact. The graphite column had longitudinal cracks and the top segment of this column showed a pale grey-brown deposit with a clear lower demarcation line. At the lower end of the column the subchannels were almost completely blocked by a mass of coarse-grained greyish matter, subsequently identified as cladding corrosion products. All seven of the uppermost fuel elements in this assembly were severely corroded at a point coincident with the lower demarcation line of the deposit on the graphite column. At this point the magnesium cladding, including the fins, was eroded down to the uranium over a distance of about 6mm. The cladding fins showed heavy surface corrosion 100mm above and below this area. These features were subsequently attributed to water flooding of the fuel assembly, with the local area of severe corrosion at the water/gas interface.

*Damaged fuel assembly No 59.*

The pressure tube and calandria tube of fuel assembly No 56 showed damage attributable to localized heating to high temperatures over a very short time. The pressure tube had ballooned locally and had a number of cracks including one through-wall crack of 10mm² cross sectional area. This damage was consistent with the impingement of a jet of incandescent matter ejected from fuel assembly No 59.

## Reconstruction of the accident sequence

In the course of the work on the blower seals, starting towards the end of October 1968, it appears that water moved past the rotating seal and entered the

*Close up view of the rupture in fuel assembly No 59.*

primary coolant system piping. During tests on 11 December 1968, in the course of which one blower was operated, the water present in the plumbing moved into some fuel assemblies, mainly Nos 49 and 59. The presence of water gave rise to serious cladding corrosion. At the water surface particularly serious corrosion took place, exposing the uranium metal. The corrosion products sank to the bottom of the pressure tube.

On 17 January 1969, in preparation for reactor restart, both main circulating blowers were turned on and the gas flow drove out the water and some of the corrosion products. However, particularly in fuel assembly No 59, sufficient corrosion products remained at the bottom of the fuel assembly seriously to restrict coolant flow in some of the subchannels. It was estimated, for example, that in the central subchannel of assembly No 59, flow was reduced to 15 per cent of its nominal value.

At about midday on 21 January, some eight hours after start up of the reactor, one or more subchannels in assembly No 59 must have been operating at higher than their rated value. However, the net effect of this would have been insufficient to give rise to a noticeable indication on the individual assembly temperature monitoring system. In the course of the power increase from 9MWt to 12MWt (17:03 to 17:15) the cladding melting point (640°C) was reached in subchannel No 7 at the point of maximum corrosion. The magnesium cladding began to melt and then flowed down the subchannel into a cooler region where it solidified, blocking coolant flow. This had two effects: first, with no gas flow up this subchannel to transport fission products, coolant activity monitoring did not detect the situation; and second, all fuel elements in the subchannel experienced a steady temperature increase so that within about 90 seconds the uranium began to melt.

About one minute later molten uranium began to flow down the subchannel. The upper elements in this subchannel steadily lost their magnesium cladding, slipped downwards and began melting. A column of molten material formed in the central subchannel (No 7), the lower portion being uranium and the upper the

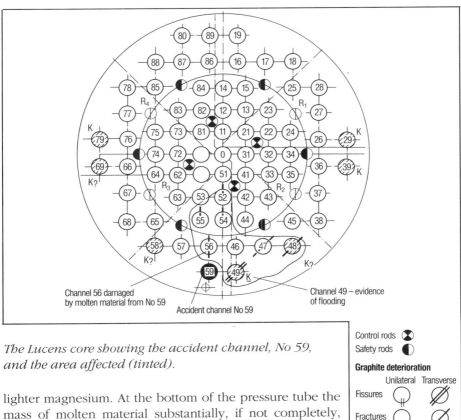

*The Lucens core showing the accident channel, No 59, and the area affected (tinted).*

Channel 56 damaged by molten material from No 59

Accident channel No 59

Channel 49 – evidence of flooding

Control rods
Safety rods

**Graphite deterioration**

| | Unilateral | Transverse |
|---|---|---|
| Fissures | | |
| Fractures | | |

lighter magnesium. At the bottom of the pressure tube the mass of molten material substantially, if not completely, obstructed the flow of coolant up the six peripheral subchannels. Clad melting, followed by fuel melting, then began in subchannel No 4 (probably the most obstructed) starting, as with the central subchannel, at the point of maximum corrosion of the cladding. Subchannels Nos 5 and 6 followed suit.

About six minutes after the melting process began a fuel element in subchannel No 1 began to burn at the point of maximum cladding corrosion. The effect of the burning in subchannel No 1 was to distort the graphite column to the extent that the mid-point of the column contacted the inner wall of the pressure tube causing rapid local heating of the tube and subsequent tube failure. It was just before this moment that the massive fission product releases to the coolant from the burning fuel triggered the shutdown system. The flow of high pressure gas into the calandria caused a rapid increase in calandria pressure. Ten milliseconds after the initial tube failure pressure reached $73 \text{lb/in}^2$ at which time the calandria tubes collapsed on to their pressure tubes, somewhat reducing the pressurization rate. As was mentioned earlier, the calandria tubes enclosing the shutoff rods were designed to resist pressures likely to be encountered in the course of an in-core LOCA and were undamaged. Also undamaged were the calandria tubes enclosing the four control rods and the four "spare" channels. At 100 milliseconds, pressure reached $117 lb/in^2$ and one of the five calandria rupture discs failed and about 1100kg of $D_2O$ was expelled into the space between the calandria and the reactor shielding.

With the failure of the pressure tube and its surrounding calandria tube, the graphite column burst under internal pressure. Graphite debris and the uppermost elements in subchannels Nos 2, 6 and 3 were ejected into the calandria through the long axial tear, leaving the upper portion of the pressure tube empty. In the lower portion of the tube, when the graphite column burst, molten uranium from the peripheral subchannels hit the pressure tube wall opposite subchannel No 4. Within about one second the pressure tube deformed locally, an oval hole was melted through the tube wall and a stream of molten metal was ejected into the calandria. This jet hit the calandria tube in position No 56, melting the calandria tube and locally heating the associated pressure tube to about 1100°C and causing it to balloon and crack in this region.

Metal water reactions from the jet of molten material and from the molten (aluminium) calandria tube produced large amounts of steam and hydrogen and caused a second rapid pressure rise in the calandria to a peak estimated to be between 235 and $350 lb/in^2$. This collapsed the calandria tubes on the control rods and spare channels, burst the remaining four calandria rupture discs and discharged more $D_2O$ from the calandria. As in the earlier pressure pulse, the calandria tubes enclosing the shut-off rods remained undamaged.

## Discussion

In their report the investigating Commission drew attention to the fact that the possibility of an accident of this type (in-core LOCA) had been recognized at the design stage, had been analyzed and the reactor had been designed to accommodate it. Attention had been paid to the overpressure implications of such an event through providing pressure relief via the calandria rupture discs as well as protection for the shut-off rods by reinforced calandria tubes in the shut-off rod positions. The safety systems called upon in the course of the accident worked as designed, both the shut-down system and containment isolation. The operators' role at this time was limited to shutting down the turbine and confirming that the reactor was safely shut down, fuel temperature was falling and the ventilation system isolated.

In order of importance the operators decided: to stop main ventilation of the reactor vault; to send a monitoring team on to the hill down wind of the plant; to contact two members of the staff who had not confirmed leaving the plant in the course of the day and to check that they were not inside the reactor vault; not to open the release valves to the rock; to wear masks in the control building; to inform the federal safety authorities; to try and complete the over-complicated form on meteorological and radiological information due to be sent to the Swiss meteorological services; to alert the police co-ordination centre of the State of Vaud; to dump the $D_2O$ and save it; to stop automatic emergency $CO_2$ feed; and to monitor the status of all alarms and controls in the control room.

The reactor safety systems all operated as expected. Detection of the damaged fuel by the coolant activity monitoring systems was certainly delayed, due to subchannel flow blockage, until a fuel element actually caught fire. However, had shutdown on the high activity parameter not occurred the reactor would still have tripped on any one of a number of other parameters shortly after the pressure tube failure, and the shut-off rod action would have been unimpeded.

*Fuel design.* The Commission's Report drew particular attention to the role in the accident played by the design of the fuel assembly. It was noted that with seven parallel subchannels, blockage of a single subchannel could have comparatively little impact on temperature or flow through the assembly as a whole. This would be especially the case for assemblies located about the perimeter of the reactor where the lower power of the fuel in this location would dictate lower flows. Thus, while the Lucens reactor was provided with quite comprehensive monitoring systems they were not able to provide early warning of subchannel flow impairment — a fact which was recognized in the plant's safety report, which, for the case of a subchannel obstruction, predicted a sequence of events essentially similar to that which actually took place. The Commission's report noted that the fuel assembly design "has not proved successful" and that while its disadvantages were recognised at the outset "it also had advantages and one of the objectives of the experimental station was to test it".

*Blower seals.* Unsatisfactory blower seal performance was a crucial contributor to the accident. Through commissioning and early operation the seals were a repeated source of trouble, with the need to prevent water ingress to the primary circuit necessitating numerous shutdowns.

The original rotating seals were well-designed according to the 1960 state of the art. They had been long-term and dynamically tested on a rig simulating normal and emergency conditions. Unfortunately, the workman capable of producing the required surface having moved on and his firm having been absorbed in a bigger company, by the summer of 1968 the operator's supplies had run out.

The research division of the reactor designer meanwhile felt that humidity in the primary circuit was not a problem and concentrated instead on maintaining a certain minimum level of humidity. As a result they specified in the summer of 1968 more modern hard-metal rotating rings with an in-leakage of water an order of magnitude too high for use in Lucens. Because the blowers had not been accepted by the operator as commissioned, due to a number of problems, the operator was not involved in the design of the new rotating rings. Had the operator been informed or consulted the problems may not have arisen.

From the end of October 1968 until the end of the year numerous adjustments and modifications were performed on the rotating graphite-metal seals, which were then tested on the installation itself since the special test rig built for seal development and testing was no longer available.

It was known that water entered the primary circuit from time to time, but it was regarded as an operating nuisance rather than a potential safety problem — all involved felt the water could not enter the fuel assemblies. In fact, it appears that no water did enter the assemblies until the blower test on 11 December 1968 drove it from its original location in the out-of-core plumbing. This moisture was the primary cause of the fuel cladding corrosion.

The humidity sensors within the primary circuit only gave useable readings above 220°C. During the cold shut down period from October 1968 to January 1969, the humidity inside the primary circuit was therefore carefully monitored day and night seven days a week by using a new type of psychrometer and by freezing and weighing $2m^3$ of the primary circuit gas content. Purging was performed at the coldest point on the primary circuit but no water condensed. Dry purging gas was also introduced at the opposite side of the primary circuit and at the labyrinth gland of the blowers.

*Condensation.* General fuel assembly clad temperatures were estimated to be about the same as the homogeneous temperatures measured on the cans of the five instrumented fuel assemblies. These temperatures were such that condensation was not anticipated. However, it is possible that peripheral fuel assemblies were cooler and water could have condensed on these assemblies.

*Viewing equipment.* Viewing equipment might have enabled the early detection of the pale grey-brown deposit found on the graphite columns of the peripheral fuel assemblies which was felt to be indicative of the presence of moisture; and indeed, the operator had asked during construction for a periscope to be installed on the fuel transfer cask inside the fuel vault so that the outer surface of the graphite blocks could be visually inspected. However, because of the additional cost (about $10 000 at that time), the request was refused.

*Containment leak.* While the containment impairment in the reactor vault had no significant impact on the offsite radiological impact (the radioactive material leaking from the loose cable penetration finished up in the machine vault) it is of interest that the cable penetration at which the major leak occurred had been improperly installed.

Two integral containment leak tests had been performed during commissioning. During the first, cable penetrations were replaced by steel plates. During the second, a cable penetration had been detected as leaking because it was badly fitted. It was rapidly repaired, but the work order did not follow the normal repair procedures. As a result, the possibility that this could be a generic failure was not identified and there was no systematic follow up. This underlines the usefulness of complete quality assurance procedures.

Individual penetration leak tests were in principle possible, but access was difficult and in any event the activities could have affected other safety-related equipment. Integral containment leak tests were anyway specified at regular intervals.

*Designed to cope.* At Lucens, therefore, an inherent design vulnerability had been recognised and factored into the safety analysis. The consequences of subchannel flow blockage were accurately predicted, even though the particular flow blockage mechanism in this case was totally unanticipated. The problems with the seal supplier are very relevant today in a period when many small, specialised companies are leaving the nuclear business. The systematic feedback of information about repair work in order to identify generic faults is also of crucial importance as plants age.

The reactor was designed to cope with this accident — but as the operator EOS now says, "who would build such a nightmare of tubing and such an inflammable fuel element again?"

## Bibliography

Commission for the Technical Investigation of the Lucens Incident, *Final Report on the Accident at The Lucens Experimental Nuclear Generating Station on 21 January 1969*, English translation 1981.

Personal communications from l'Energie de l'Ouest-Suisse.

See also: IAEA -SM 234/8 on decommissioning the Lucens reactor plus papers delivered to the 1982 Seattle Symposium on Decommissioning.

# THREE MILE ISLAND

The Unit 2 reactor at Three Mile Island was a Babcock and Wilcox designed 959MWe (880MWe net) pressurized water reactor operated by Metropolitan Edison. The unit achieved first criticality on 28 March 1978 and, precisely one year later, on 28 March 1979, the reactor core was destroyed when, following a total loss of feedwater, core cooling was seriously impaired.

## Plant description

The reactor core comprised a 3.27m diameter by 3.65m high cylinder made up of 177 fuel assemblies, contained in a 4.35m diameter by 12.4m high carbon steel pressure vessel. Each fuel assembly was a 15x15 array containing 208 fuel rods plus spaces for control and instrumentation elements. The fuel was zircaloy-4 clad uranium dioxide enriched to 2.57 per cent U-235.

The reactor's primary coolant circuit comprised two loops, each with two main circulating pumps and a single vertical once-through steam generator. Primary coolant operating pressure was 2150lb/in$^2$ with an outlet temperature of 319.4°C. A pressurizer connected to one of the loops controlled primary system pressure and provided surge volume to accommodate expansion and contraction of primary coolant water. The pressurizer was connected to a reactor coolant drain tank via a power operated relief valve (PORV). A block valve was provided upstream of the relief valve and could be remotely operated in the event that the relief valve stuck open or leaked.

*The Three Mile Island station. Unit 2 is on the right.*

Under normal operating conditions a primary coolant let-down and make-up system removed between 45-70 USgpm from the primary system, purified and cooled it and returned it to the primary system through one of three high pressure make-up pumps. Make-up supply was automatically controlled by pressurizer level signals. Two of the make-up pumps formed part of the emergency coolant injection system. When actuated each pump could inject up to 300 USgpm per pump of borated water at 2400lb/in² into all four reactor cooling inlet pipes. Under injection conditions the let-down flow was automatically stopped.

Feedwater flow was from the condensate pumps through eight parallel resin beds (condensate "polishers") to condensate booster pumps, through low pressure feedwater heaters to the main feedwater pumps (one per steam generator) and through the high-pressure feedwater heaters to the steam generators. The main feedwater pumps were backed up by two electric (DC) pumps (one per steam generator) and a single, shared, steam turbine driven pump. On loss of main feedwater pumps these auxilliary pumps would start automatically and provide a direct feed to the steam generators. The valves linking these pumps to the steam generators were arranged to open when steam generator level fell to 76.2cm or less.

The reactor, primary circuit pumps and piping and steam generators were enclosed in a pre-stressed post-tensioned concrete containment structure lined with steel 1.22m thick.

*Keeping an eye on defuelling operations in the Three Mile Island 2 clean-up command centre.*

## Accident sequence

At 04:00 on 28 March the reactor was running at 98 per cent of full power. Three members of the operating staff were working on the No 7 condensate polisher, breaking up the compacted resin bed with compressed air and water before transferring the resin to the regeneration system. At this moment all the in-service condensate polishers isolated automatically as a result of water entering an instrument air line through a check valve which had stuck open. This interruption in condensate flow immediately caused the condensate booster pumps and main feedwater pumps to trip. The turbine tripped in turn and the reactor control system automatically began reducing reactor power. The loss of feedwater pumps activated the three auxiliary feedwater pumps.

Within three seconds primary system pressure had risen to the pressurizer relief valve setpoint (2255lb/in$^2$g) and this valve opened. At 7.2 seconds the primary system pressure trip setpoint was reached (2355lb/in$^2$g) and the reactor tripped automatically. The operators saw the pressurizer level falling (at this point the rate of heat removal by the steam generators was greater than the decay heat production rate) so they stopped the letdown flow and unsuccessfully attempted to start a second high pressure makeup pump. The attempt was unsuccessful because the operator did not hold the startup switch closed for long enough for the lubricating oil pump to build up sufficient oil pressure to permit pump startup.

By 12 seconds, primary system pressure had fallen to 2205lb/in$^2$g, at which the pressurizer relief valve should have closed. It did not do so, although an indicator light in the control room seemed to suggest that it had. In fact, all the light indicated was that the signal to open the valve was no longer in the circuit — there was no direct indication of actual valve status.

With this valve open, primary coolant continued to discharge from the system at an initial rate of about 200 USgpm.

After 30 seconds steam generator level had fallen to 76.2cm — the point at which the control valves admitting auxilliary feedwater flow opened. However, the block valves in the auxilliary feedwater system had been left closed and no water reached the steam generators. With no secondary side heat sink the reactor primary coolant system began to heat up. At about this time the second high pressure make-up pump was successfully started. The combined effects of two make-up pumps operating, no let-down flow and heat-up of the primary system due to the deteriorating heat sink provided by the rapidly drying-out steam generators, caused pressurizer level, which had fallen to 401.32cm, to begin to rise.

Reactor primary system pressure reached 1640lb/in$^2$g at 121 seconds, at which point the high pressure injection system started up automatically. This involved automatic startup of the third high-pressure make-up pump (pump C), and full opening of all make-up pump discharge valves. High pressure make-up pump B tripped off, leaving pumps A and C supplying injection cooling water at the maximum rate of 600 USgpm. However, the pressurizer level continued to rise.

The operators, after taking over manual control of the pumps, began throttling pump injection flow after about two and a half minutes' operation in an effort to halt the pressurizer level rise and, shortly afterwards, tripped one injection pump while the other continued to run throttled to about 100 USgpm. In addition, let-down flow was initiated at its maximum rate of 140 USgpm. These actions had the effect of causing pressurizer level to fall for about 15 seconds, after which it began to rise again and the pressurizer bubble was lost. At this point the reactor primary system pressure had reached saturation for a large volume of coolant. From then on the reactor outlet (hot leg) temperature remained at saturation as determined by the saturation pressure.

At about the time the steam generators boiled dry (120 seconds), the temperature of the water in the reactor coolant drain tank (to which the relief valve discharged) showed a significant increase. However, the meter displaying this information was located on the back of the main control panel and was not visible from the normal operating location. Even had it been observed, however, the reading need not necessarily have lead to the inference of an open relief valve since the liquid in the drain tank was already warm, due to chronic pressurizer leakage, and would have been given additional heating from the initial release.

Two minutes later the drain tank pressure rose rapidly to $120lb/in^2g$ and oscillated about this level. At the time when the pressurizer bubble was lost (six minutes) the drain tank pressure rose rapidly to $150lb/in^2g$, consistent with the delivery of water rather than steam from the pressurizer. Water discharged from the drain tank flowed to the reactor building sump and a sump pump automatically started some two minutes later. This pump should have discharged to a waste hold-up tank in the auxilliary building but instead was directed to the auxilliary building sump tank which was not only already full, but also had a blown rupture disc (scheduled for later repair).

At eight minutes an operator checked the auxilliary feedwater valve line-up, discovered the closed block valves and immediately opened them. Steam generator water levels did not recover immediately since the high primary system temperature and the heat stored in the steam generator hardware caused immediate flashing, but over the next 15 minutes primary system temperature decreased. However, because of the outflow through the open relief valve, and resulting depressurization, the system continued to operate at saturation temperature.

At 15 minutes the rupture disc in the reactor coolant drain tank, to which the pressurizer relief valve discharged, burst at $192lb/in^2g$ and drain tank pressure fell rapidly to $10lb/in^2g$. In response the reactor containment building pressure rose by $1lb/in^2g$, and, about four minutes later, the reactor building exhaust air monitor showed a factor of ten increase, reflecting releases of the normal radioactive gas inventory from the reactor coolant.

It was at about this time that the gradually increasing void fraction in the core was indicated by a rising signal from the source range neutron detector. Pump performance also began to degrade and the first pump overspeed alarms were received.

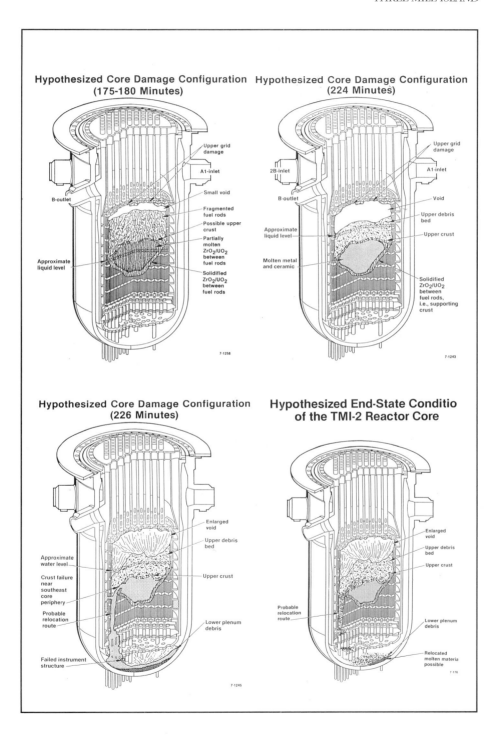

**Hypothesized Core Damage Configuration (175-180 Minutes)**

Upper grid damage
A1-inlet
Small void
B-outlet
Fragmented fuel rods
Possible upper crust
Partially molten $ZrO_2/UO_2$ between fuel rods
Approximate liquid level
Solidified $ZrO_2/UO_2$ between fuel rods

7-1258

**Hypothesized Core Damage Configuration (224 Minutes)**

Upper grid damage
2B-inlet
A1-inlet
Void
B-outlet
Upper debris bed
Approximate liquid level
Upper crust
Molten metal and ceramic
Solidified $ZrO_2/UO_2$ between fuel rods, i.e., supporting crust

7-1243

**Hypothesized Core Damage Configuration (226 Minutes)**

Enlarged void
Upper debris bed
Approximate water level
Crust failure near southeast core periphery
Upper crust
Probable relocation route
Lower plenum debris
Failed instrument structure

7-1245

**Hypothesized End-State Conditio of the TMI-2 Reactor Core**

Enlarged void
Upper debris bed
Upper crust
Probable relocation route
Lower plenum debris
Relocated molten materia possible

7-176

73

After 25 minutes the primary system pressure was approaching the pressure in the secondary system and the resulting reduction in heat transfer via the steam generators allowed steam generator water levels to approach their control setpoint. Secondary system pressure, determined by the turbine by-pass system, was about $1025lb/in^2g$ and primary system pressure remained slightly above this. Thus, at this point the principal decay heat removal mechanisms were boil-off of primary coolant, with release of a two-phase mixture from the pressurizer relief valve, plus a small contribution from the steam generators.

At 73 minutes both main circulating pumps in the B coolant loop were turned off, an action which was in accordance with the station's operating procedures given the pertaining reactor coolant system temperature and pressure conditions. The rationale for such a procedure is protection of the pumps and pump seals.

By now the situation was:

❑ The pressurizer relief valve was still open.
❑ Coolant flow had stopped in the B loop and was decreasing steadily in the A loop.
❑ Reactor coolant pressure was at about $1000lb/in^2g$ with inlet and outlet temperatures in the range of $287.8°C$.
❑ Neutron flux levels were increasing.
❑ Reactor building pressure was increasing steadily.
❑ The reactor building sump was overflowing.
❑ Increasing radiation levels were being signalled at various locations in the plant.
❑ Pressurizer level was indicating at the high end of the scale and remaining relatively constant.
❑ A letdown flow of 100 USgpm was being maintained in an effort to regain a steam bubble in the pressurizer.

Under normal conditions, turning off the B loop pumps would result in flow reversal through this loop due to the continued operation of the pumps in the other loop. However, phase separation occurred in the B loop, and steam in the inverted "U" section of the loop piping blocked flow through the loop.

In the A loop the pumps continued to circulate a two-phase mixture of continuously increasing void fraction while loop B remained stagnant. Heat was now being transferred from the core by, first, evaporation of primary coolant and release of a two-phase mixture from the pressurizer and the letdown line, and, second, transfer via the A loop steam generator.

The flowrate in the A loop continued to decrease due to increasing void formation and consequently decreasing pumping effectiveness. As with the B loop pumps, in accordance with operating principles both A loop pumps were turned off one hour 40 minutes after turbine trip. Less than 20 seconds after the pumps were switched off primary coolant flow dropped to zero. For about 90 seconds out-of-core neutron instrumentation indicated a decreasing flux level, attributable to phase separation. Since there was sufficient coolant inventory at that point to fill the downcomer annulus and to cover the core the source range monitors sensed a normal thermalhydraulic condition. However, two minutes later measured flux levels began rising again as water boiled away and began to uncover the core. About 10 minutes after the A loop pumps were turned off the hot leg temperatures in both loops began to rise until they went off scale high (326.7°C) about 19 minutes later, while the cold leg temperatures went off scale low. The upper part of the core, now dry, had begun generating superheated steam.

At this point the primary system pressure began to rise and the reactor building radiation readings increased rapidly until they went off scale high. The increase in pressure was due to the evolution of hydrogen from the zirconium-water reaction between the fuel cladding and steam and the increased radiation readings were due to fission products which, released from the damaged fuel, had reached the reactor building via the pressurizer relief valve and the drain tank.

Two hours and 22 minutes after turbine trip the stuck open relief valve was discovered and an upstream block valve was closed, halting the escape of primary coolant from the system. Primary system pressure began to increase. In an attempt to hasten primary system pressure recovery an additional high-pressure make-up pump was started (at two hours 34 minutes) and run for 12 minutes after which it was turned off. Attempts were also made to start a primary coolant pump. Neither of the A loop pumps could be started but the second pump in the B loop was and some coolant flow in the B loop was established for a few seconds, then fell off to zero. The pump continued to run (with a high vibration alarm) for about 18 minutes.

At the time this pump was started the upper region of the core was in a highly oxidised and brittle condition and a pool of molten fuel and cladding was forming in the central portion of the core. The water level was about two feet above the bottom of the core. Water delivered by the B loop pump shattered the fuel in the upper portion of the core, resulting in the formation of a layer of debris with a void above. A mixture of steam and hydrogen now filled the upper portions of the reactor vessel and the primary heat transport system piping, impeding the re-establishment of primary system flow.

At two hours 55 minutes into the accident it had become clear that plant conditions had gone beyond those catered for in the station emergency procedures and a site emergency was declared followed by a general emergency (including notification of offsite authorities) 26 minutes later.

Coolant flow could not be established using the main circulating pumps so attempts were made to provide cooling by a "feed and bleed" approach. The pressurizer block valve was opened at the time the B loop pump was stopped (three hours and 12 minutes) and closed five minutes later with primary system pressure at about 1925lb/in²g. One of the high pressure injection pumps was started and run for 17 minutes during which time primary system pressure fell to 1450lb/in²g, as a result of condensation of steam by the cool injection water. The block valve was then re-opened. About five minutes later there was an increase in the count rate registered by the out-of-core flux detectors and reactor coolant pressure rose to 1675lb/in$^2$g as a result of what has subsequently been determined to be the relocation of several tonnes of molten fuel from the central region of the core to the lower head of the reactor vessel.

At four hours 27 minutes a period of sustained high pressure injection was begun using two injection pumps. Primary coolant system pressure rose, indicating that injection water was now filling the reactor vessel rather than just condensing steam.

Following closure of the relief block valve at five hours 18 minutes, system pressure rose and was maintained at 2000 to 2200lb/in$^2$ through subsequent cycling of this valve. About two hours later (seven hours 38 minutes) injection was halted and the relief block valve was opened in an attempt to depressurize the system to allow injection of water from the core flood tanks. System pressure dropped to about 435lb/in$^2$g over the next 90 minutes, initiating a relatively small injection of water from the core flood system (about 1540 USg) and the relief valve was closed.

Discharge to containment through the relief valve included significant amounts of hydrogen and at nine hours 49 minutes (about 40 minutes after the last closing of the relief valve) a hydrogen burn in the containment building caused a pressure spike of 28lb/in$^2$g, activating high-pressure injection, reactor building isolation, reactor building sprays and decay heat removal pumps.

At 10 hours the relief valve was again opened and primary system pressure fell to a low of 410lb/in$^2$g. Again, little water (about 320 USg) was injected. Pressure

*Health physics technicians working inside the Three Mile Island 2 containment during the post accident clean up operations.*

was maintained at about $420lb/in^2g$ (using a single injection pump) until closure of the relief valve 68 minutes later (11 hours eight minutes).

With the relief valve closed pressure gradually rose to $620lb/in^2g$. For about four hours there was minimal make-up flow, no heat transfer through the steam generators and only occasional flow through the relief valve (the valve was opened twice for a total time of 19 minutes over this period) with the result that the water in the core region was slowly boiling and the water level falling. This would seem to be confirmed by the fact that when high pressure injection was resumed (at 13 hours 23 minutes) for the first eight minutes system pressure fell as did pressurizer level, indicating that the inflow of water was condensing steam.

High pressure injection was shut off after 80 minutes with the system substantially filled, the remaining pockets of non-condensible gas being compressed. However, the compressed steam/hydrogen mixture still blocked natural circulation, despite an attempt to initiate it through briefly running one of the main circulating pumps. At 15 hours 50 minutes this circulating pump was started and allowed to continue running and forced circulation was achieved in the "A" loop. Primary system pressure stabilized at 1000 to $1100lb/in^2g$ and temperatures indicated a cooling trend. Essentially this marks the end of the accident sequence.

## Discussion

An immediate reaction to the Three Mile Island accident must be an appreciation of the very rugged design of the reactor and its containment structure. Despite grievous core damage, including the melting of a significant proportion of the fuel, radioactive releases were low enough that, according to the Kemeney Commission Report, the theoretically most exposed offsite individual would have received only 70 millirem. The same source suggests a collective dose for the population living within a 50 mile radius of the plant of 2000 person rem, about 0.8 per cent of that due to natural background radiation.

The Three Mile Island accident is invariably described in terms of minor failures followed by inappropriate operator response — or "mechanical failure" and "operator error". The crucial mechanical failure was the failure of the PORV to close and the crucial "operator errors" were throttling the high pressure injection flow and turning off the main circulating pumps.

The term "operator error" seems to be both unfair and misleading in this context. In the first few minutes of the accident the operators were faced with what appeared to be a straightforward trip following loss of feedwater, and they followed the established plant procedures for reactor trip recovery which included a requirement for the maintenance of a specified pressurizer level (254cm).

Additionally, another Unit 2 Operating Procedure stated: "The pressurizer must not be filled with water to indicated solid water conditions (1016cm) at any time except as required for system hydrostatic tests" and operator training laid especial emphasis on the necessity of avoiding a solid system. It should be noted that the simulator model used in operator training failed when the pressurizer filled.

It also seems unreasonable to argue that the operators should have been able to infer that their plant was suffering from an unanalysed event — a small break loss of coolant accident with the leak path through the pressurizer relief valve — and act accordingly. Twelve seconds after the reactor trip the PORV was indicated "closed". A high pressurizer relief valve outlet temperature alarm was transmitted to the alarm printer in the control room 30 seconds after reactor trip but was not printed out for several minutes because the printer was then receiving alarms at the rate of over 100 per minute and had become overloaded.

Three minutes and 13 seconds after the trip the reactor coolant drain tank relief valve lifted and 13 seconds later a drain tank high temperature alarm was transmitted, events which could have been construed as indicating a continuing steam-water flow from the pressurizer. However, it is scarcely suprising that these, as well as subsequent reactor coolant drain tank signals, did not alert the operators — drain tank parameters were not shown on the panels in the immediate view of the operators, but were behind the principal control panels. In order to pick up the relevant alarm annunciation the operators would have first had to clear all the audible alarms on the front panels — they were all sounding at once.

It is clear that from the moment of the reactor trip onwards the operators were having to function in a high stress environment with a large number of demands on their attention. For example, the condenser hotwell high level alarm point was reached at 73 seconds — the airline to the level controller was broken in the initial transient — and prompted unsuccessful attempts to regain hotwell level control. This was of considerable concern to the operators since failure to correct the situation could lead to loss of condenser vacuum which would require steam dumping to atmosphere.

Another example of a diversionary demand upon the operators was the concern about recriticality. About 20 minutes after trip the increase in neutron flux signals (resulting from increasing steam voids in the core) prompted the operators to push the reactor trip button and, subsequently, to investigate the possibility of a demineralized (ie unborated) water flowpath into the make-up system.

While it becomes quite clear from such detailed *post-facto* sequences of events as that compiled by the Nuclear Safety Analysis Centre that the Three Mile Island reactor was suffering from a small break loss of coolant accident, and that early operator action could have completely averted reactor damage, this was by no means clear to the operators at the time. Much discussion of this accident has been couched in a form from which it is possible to infer a lack of perspicacity on the part of the operating crew. Not only is this grossly unfair to the people concerned, but it can lead to a serious misinterpretation of the causative factors. The situation has some similarity with the Chernobyl accident in that operator actions play a crucial role, but, as with Chernobyl, the operator actions are of less interest themselves than what gave rise to those actions.

Central to the Three Mile Island accident is the failure of the pressurizer relief valve to close. This event had not been identified as a small break loss of coolant accident in the safety analysis. Nor had its significance been formally reflected in operator training or operating procedures.

The Kemeny Report draws attention to the fact that in 1977, at a Babcock and Wilcox plant similar in design to Three Mile Island (Davis Besse), a PORV jammed open and operators responded to the resultant rising pressurizer level by throttling water injection. The plant was at low power (9 per cent) at the time, and the block valve was closed after about 20 minutes. No fuel damage resulted. Both the Nuclear Regulatory Commission and Babcock and Wilcox investigated the incident and a B&W engineer subsequently noted in an internal memorandum that if Davis Besse had been operating at full power at the time of the incident "it is quite possible, perhaps probable, that core uncovery [sic] and possible fuel damage would have occurred". In addition, Kemeny notes, in January 1978 an NRC official pointed out the likelihood of "erroneous operator action" in a TMI type event and a Tennesse Valley Authority engineer analyzed the problem of rising pressurizer level and falling pressure. The latter analysis was given to B&W, the NRC and the Advisory Committee on Reactor Safeguards. No information calling attention to the possibility of inappropriate operator response to an open PORV or the implications of such a response was provided to utilities by either B&W or the NRC before the Three Mile Island accident.

# Bibliography

John G. Kemeny, et al, *Report of the President's Commission on the Accident at Three Mile Island*, US Government Printing Office, Washington, 1979.

Mitchell Rogovin and George T Frampton, *Three Mile Island: A Report to the (NRC) Commissioners and to the Public*, NUREG/CR-1250, 1980.

*Analysis of the Three Mile Island Unit 2 Accident NSAC-80-1*, Nuclear Safety Analysis Centre, California, 1980.

*Proceedings of the First International Information Meeting on the TMI-2 Accident*, CONF-8510186, EG&G Idaho Inc and US Department of Energy, October 1985.

# CHERNOBYL

The Chernobyl nuclear power station is located in the Soviet Union about 100km north of the city of Kiev. At the time of the accident the station comprised four RBMK 1000 reactors in service and two under construction (now abandoned). Unit 4 came into service in late 1983 and was destroyed on 26 April 1986 as a result of a power transient triggered in the course of a safety system test.

## Reactor description

The RBMK reactor is a vertically orientated, graphite moderated, direct cycle (boiling water) pressure tube reactor. It essentially comprises a 12m diameter by 7m high cylinder built up of graphite blocks, threaded by almost 1700 vertical zirconium alloy (Zr-2.5 wt per cent Nb) pressure tubes which contain the reactor fuel and the various control and shut-off rods. The reactor fuel is zirconium alloy clad uranium oxide enriched to 2.0 per cent U-235.

The heat transport system comprises two cooling loops, each with four main circulating pumps (three running and one on standby) and two steam drum/separators. The discharge from the pumps goes to a large (900mm diameter) header which feeds a series of 300mm diameter distribution headers which serve groups of feeder pipes. As the coolant flows up the fuel channels it boils and the steam/water mixture, with an average quality of 14 per cent (maximum 22 per cent), is fed to the steam

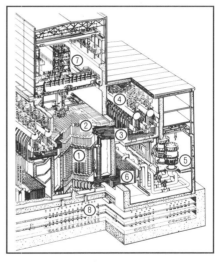

*Cutaway of the Chernobyl 4 plant. (Source: KWU/IAEA)*

**Key:** 1 – core. 2 – upper shield and pressure tube closures. 3 – steam/water piping. 4 – steam drum/separator. 5 – main coolant pump. 6 – water inlet piping. 7 – refuelling machine. 8 – pressure suppression pool.

drum, thence to the two 500 MWe turbines. Coolant flow through the core is matched to power levels by large throttling valves on the main coolant pump discharges. In addition, at the point of each feeder connection to a distribution header, there is an individual control valve which is used to regulate the flow through the individual channels to maximize the margin to fuel dryout under a variety of operating conditions while keeping the channel power as high as possible.

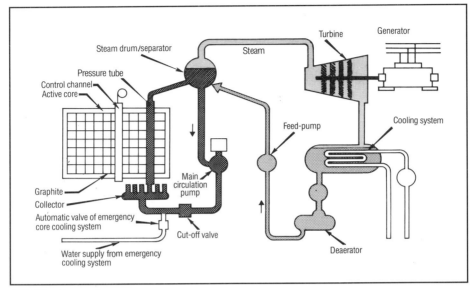

*RBMK schematic.*

The reactor core is contained in a steel vessel which is filled with a mixture of helium and nitrogen to maintain an inert atmosphere for the hot graphite (about 600°C) and to promote heat transfer. An arrangement of graphite rings on the pressure tubes ensures good thermal contact between the tubes and the moderator to provide a heat transfer path from the graphite via the pressure tubes to the coolant.

| Table of rod types | |
| --- | --- |
| Manual control | 139 |
| Local automatic regulation (LAR) | 12 |
| Automatic power regulation | 12 |
| Emergency protection (AZ) | 24 |
| Bottom entry short absorbers (axial shaping) | 24 |
| Auxiliary absorbers (temporarily installed to hold down initial reactivity, progressively replaced by fuel) | 240 |

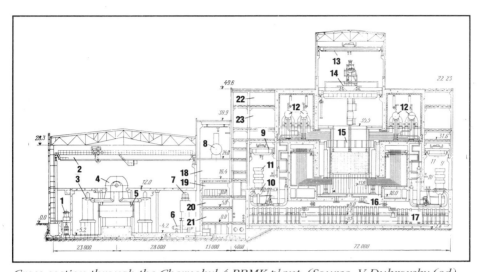

*Cross-section through the Chernobyl 4 RBMK plant. (Source: V.Dubrovsky (ed), Construction of Nuclear Power Plants, Mir Publishers, Moscow, 1981)*

**Key:** 1 – first stage condensate pump. 2 – 125/20t overhead travelling crane. 3 – separator steam superheater. 4 – K-500–65/3000 steam turbine. 5 – condenser. 6 – additional cooler. 7 – low pressure heater. 8 – deaerator. 9 – 50/10t overhead travelling crane. 10 – main circulating pump. 11 – electric motor of main circulating pump. 12 – steam drum/separator. 13 – 50/10t remotely controlled overhead travelling crane. 14 – refuelling machine 15 – core. 16 – accident containment valves. 17 – pressure suppression pool. 18 – pipe aisle. 19 – modular control board. 20 – location beneath control board room. 21 – house switchgear locations. 22 – exhaust ventilation plant locations. 23 – plenum ventilation plant locations.

Reactor control is via a total of 211 boron carbide absorber rods of which 24 are designated "safety", "emergency protection", or "scram" (AZ). These rods are usually fully outside the core to provide reactivity depth on shutdown. There are also 12 "local automatic regulation" rods, which maintain power shape using signals from four lateral ionization chambers, and 12 "automatic power regulation" rods which maintain total reactor power and are arranged in three sets of four ganged rods (AR1, AR2 and AR3). The 139 "manual control" rods are manipulated by the operators in response to changing reactor conditions to keep the automatically controlled rods within their range of travel. Finally, there are 24 "bottom entry" short rods used for manual axial power shaping. Auxiliary absorbers are also installed temporarily in a new core to hold down reactivity. There are 240 of these boron steel absorbers initially, which are progressively replaced by fuel as burnup increases.

The reactor and most of the coolant circuit is contained within a number of thick-walled leak-tight enclosures which are linked to two pressure suppression pools ("bubbler ponds"), however the upper sections of the fuel channels and the steam drums are not so enclosed. The top of the reactor is covered with a 3m thick steel and concrete shielding slab through which the fuel channels pass. Fuelling is on-power, using a single fuelling machine above the reactor. This machine also serves the adjacent unit.

A significant feature of the RBMK reactor design is the influence of reactor coolant conditions on the power coefficient at various power levels. At higher power levels the positive reactivity effect of an increase in coolant void is more than offset by the negative reactivity effects of an increase in fuel temperature so power rises are self-limiting. However, at power levels below about 20 per cent of full power this is not the case and coolant void is the dominant influence on the power coefficient. At low power levels the reactor is unstable and difficult to control. For this reason a fundamental operating procedure forbids sustained operation at power levels below 20 per cent.

Another feature of interest is that the maximum rate of insertion of the rods (prior to modifications made after the accident) was 400mm/s — implying an insertion time of about 15 seconds for a rod moving from the fully withdrawn position. This relatively modest insertion speed led to another important operating procedure that required operators to maintain a "reactivity margin" of no fewer than 30 rods inserted at least 1.2m into the core at all times.

## The voltage regulator test

The test which triggered the catastrophic reactivity excursion at Chernobyl 4 was aimed at improving capabilities for coping with emergencies involving loss of essential power supplies.

The purpose was to confirm that a new voltage regulation arrangement would extend the time for which a coasting down turbine generator could be used to run the emergency core cooling system (ECCS) pump in the event of a station blackout (loss of electric power). The idea was to show that turbine generator inertia was sufficient to power the ECCS for enough time (35 seconds or so) to get the emergency diesels started. It seems that the test should have been done during the initial start-up of unit 4, but since this had coincided with New Year's Eve, the test had been postponed and the plant had entered service in breach of safety requirements.

In the test planned for 25-26 April, rather than using the ECCS pump itself, the load of the ECCS pump was to be simulated by running four main coolant pumps from the coasting down turbine generator. The test was to be performed at a level of 22-32 per cent of full reactor power (ie at 700-1000MWt). It was therefore planned to do it as the unit was coming down for a routine maintenance shut down. The planned steps were:

- ❑ Reduce reactor power to the range required for the test and shut down one of the two turbine generators.
- ❑ Switch off the ECCS to prevent it operating spuriously during the test.

❏ Reconfigure power supplies to the main coolant pumps so that four (two in each of the plant's two cooling loops) were supplied from the remaining turbine generator and the other four were connected to the grid. The plan was that at the end of the test the four pumps connected to the grid would still be operating, providing the necessary cooling to keep the unit in a safe condition.
❏ Shut off the steam supply to the turbine and measure the time for which the electrical load was maintained.

This was the planned test procedure. What actually happened was rather different.

At 01.00 on 25 April power reduction started for the routine maintenance shutdown. It was planned to carry out the voltage regulator trial at 22-32 per cent of full power (700-1000MWt).

50 per cent power was reached at 13.05 and one of the plant's two main turbines was disconnected, all house load being transferred to the remaining turbine. At about this time a request was received from the grid controller in Kiev for generation to continue at unit 4 following problems at a thermal station in the network. It became necessary to maintain the power level at 50 per cent and delay the experiment. This power was maintained for nine hours, until 23.10, when power reduction was resumed.

In accordance with the experimental programme the ECCS had been disconnected at 14.00 and remained disconnected. As it turned out this had no bearing on subsequent events, but the willingness to operate for an extended period without an ECCS is perhaps a reflection of the attitudes to safety that seemed to prevail in the run up to the accident.

At 00.28 on 26 April power reduction was proceeding as intended when a mistake in setting control rods caused power to fall to less than 30MWt (around 1 per cent of full power) — below the 700MWt that is the minimum level specified in operating procedures for continuous operation and also well below the 700-1000MWt level at which the test was to be carried out. The error — a "genuine" operator error unlike some of the deliberate violations which occurred later — happened when switching from the local to the global power regulating system. A "hold power" command was not entered, causing power to continue to fall below the intended level.* The power fell rapidly and uncontrollably to a minuscule level because of reduced boiling (ie collapse of voids) in the core — a manifestation of

---

* On current information the precise details of this error (as well as several other aspects of the accident) are difficult to pin down. According to the Soviet report presented to the August 1986 Vienna post accident meeting "when the local automatic regulation system was shut off, which under the operating rules is supposed to be done at low power, the operator was unable to eliminate the resulting imbalance in the measuring part of the automatic regulator quickly enough. As a result of this, the power fell below 30MWt."

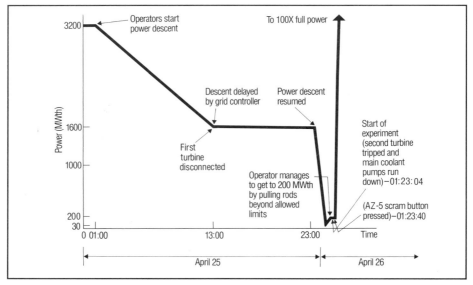

*Power levels at Chernobyl 4 in the hours preceding the accident. Power was allowed to go well below the 700–1000 MWt level at which it was originally planned to do the test.*

the RBMK's positive void coefficient. Because more water is present more neutrons are absorbed and power is reduced — the opposite of the feedback process that led to the final fatal power excursion.

## The xenon trap

Raising power from this extremely low point was exceedingly difficult, principally because of xenon poisoning. Xe-135 is produced in a reactor during normal operation through decay of the fission product I-135 (with a half life of 6.6 hours). Xenon is a powerful neutron absorber (or "poison"). In steady operation at a given reactor power an equilibrium level of Xe-135 is reached corresponding to that power level. This is because neutrons tend to "burn out" the Xe-135 (ie convert it to other, less absorbing, isotopes). This burning out process counterbalances the process of Xe-135 creation and an equilibrium value is achieved.

When power fell rapidly to 30MWt or thereabouts at Chernobyl 4, which had been at 1600MWt for an extended period, the xenon remained at its 1600MWt level. I-135 was also at its 1600MWt level, leading to the generation of additional Xe-135. The fission process was effectively smothered by the neutron hungry Xe-135, making power raising very difficult indeed.

By 01.00 on 26 April the operator had managed to get power up to 200MWt (about 7 per cent of full power). But, because of the extent of xenon poisoning the power could only be raised by manually withdrawing a large number of control rods from the core.

The withdrawal of so many rods contravened operating procedures, which specified a minimum "operating reactivity margin" (ORM) of 30 rods equivalent inserted in the core. Valery Legasov, head of the Soviet delegation, told the August 1986 post-accident meeting in Vienna that to go below 30 required the permission of the station chief engineer, but if the "margin goes down to 15 then no-one in the whole world, including the president of the country, can allow operation. The reactor must be stopped". (The operators eventually got down to 6-8 rods inserted.)

The purpose of specifying a minimum ORM is to make sure operators keep enough rods in the core to ensure that the reactor can always be shut down quickly enough in case of emergencies. Maintaining the equivalent of some minimum number of rods in the core is intended to guarantee that negative reactivity will be inserted above some minimum rate in the event of a scram.

A further problem with having such a low ORM is that it means that the amount of absorbing material in the core is reduced to a minimum, which has the effect of making the void coefficient of the reactor yet more positive.

Another major contributor to the eventual accident was that the reactor was now being operated at well below the 20 per cent power level specified as the minimum for continuous operation. At low power not only was it difficult to keep tabs on the thermal hydraulic parameters (water temperatures, flows, levels, pressures, degree of voiding etc) but also the positive void coefficient dominated the overall power coefficient of the reactor (outweighing the fuel temperature coefficient, which is negative).* This meant that the overall power coefficient was positive — ie that a rise in reactor power would lead to a further rise in power leading in turn to a further increase, a fatal feedback loop. This was why operation below 20 per cent power was strictly prohibited.

---

* The precise value of the void coefficient depends on characteristics of the core, such as initial enrichment of fuel and burn-up. It varies in the course of one operating cycle. At the start of life some channels in the core are loaded with fixed absorbers to hold down excess reactivity, and the void coefficient is negative. As the cycle continues, the fuel burns up and absorbers are taken out. This has the effect of causing the void coefficient to become increasingly positive — as it was at the time of the accident. At normal operating power levels, the void coefficient is nevertheless outweighed by the Doppler coefficient, which is always negative and reflects the fact that as fuel temperature rises, the bandwidth of neutron energies that can be absorbed by the U-238 atoms increases. However at low power the void coefficient dominates. At the time of the accident the operators had got the core into a situation where the void coefficient was at a peak level — about $+2 \times 10^{-4}$ per cent.

As if all that were not enough, the axial flux profile in the reactor was now very distorted and getting more so. As a result of xenon poisoning and withdrawal of so many control rods, the profile had taken on a double humped shape, with peaks at top and bottom of the core, the largest being at the bottom. One effect of this was to set the stage for large power increases at the bottom of the core by magnifying the effects of any voiding in the channel inlets (which are in the lower part of the core). Rods inserting from the full out position at the top of the core would not be very effective in controlling any such power rise at the bottom of the core.

Nevertheless, preparations for the test continued — despite the fact that reactor power was nowhere near that laid down in the test programme.

At 01.03 and 01.07 on 26 April, in accordance with the test programme, an additional main coolant pump was switched into each of the main cooling loops. There were thus a total of eight pumps operating. However, because the power level was much lower than anticipated in the test procedure, the hydraulic resistance was low and the flow through the core rose to an extremely high level — with some pumps exceeding their permitted flow rates by a large margin, constituting yet another violation of operating rules. The consequence was a sharp reduction in steam production, leading to a fall in pressure and a lowering of water level in the steam drum/separators. The whole circuit was close to boiling (perhaps 1 degree C away) but effectively no boiling was actually taking place. This virtual absence of voiding meant that the system was operating in a regime where it was more than usually sensitive to increases in voidage when they did occur. The operators tried to control the core flow with throttling valves but were unsuccessful.

To prevent the low water levels in the the steam drum/separator from tripping the reactor, at 01.19 on 26 April the operators disconnected the relevant channels of the protection system. In addition, because they were having trouble with system pressure they also disabled the trip on that parameter. To try to increase the levels in the steam drum/separator, the operators increased the feedwater flow. The cooler feedwater caused steam generation to drop even more, with a consequent further fall in reactivity. This prompted the operator, at about 01.19.30, to remove even more manual control rods, with a consequent further reduction in the operating reactivity margin, which was already so low as to be in gross violation of operating rules. At this point the core average void was calculated to be zero. To try and control the fall in pressure the turbine steam bypass valve (steam dump) was closed at 01.19.58.

At 01.22 on 26 April the operator decided that the level in the steam drum/separator was now high enough and abruptly reduced feedwater flow. This meant that over the next minute or so relatively warm water would begin to arrive at the channel inlets at the bottom of the core — causing void growth to begin.

At 01.22.30 the operator called up a print out of his operating reactivity margin so that he could define neutron fields and reactivity margin before starting the test. It indicated that the operating reactivity margin was down to 6-8 rods — less than half the minimum permissible under any circumstances (and a quarter of that specified for normal operation). This required immediate shutdown. But "the staff were not stopped by this and began the experiment. This situation is difficult to understand", Valery Legasov said.

By 01.23.04 the operators, surmising that the system was closer to stable than it had been for some time, decided to proceed with the test. Accordingly, the stop valve of the turbine was closed so that it would run down in accordance with the plan. Normally the reactor would have scrammed on this loss of the second turbine (the other of the unit's twin turbines having been shut down the day before in preparation for the test). But the operators had blocked this trip to allow a re-run of the test if necessary.

From about 01.23.30 reactor power began to rise (albeit slowly at first) because of increased steam production (ie voiding). The increased steam production was due to the run down of the main coolant pumps (as a result of the turbine shut off) and an increase in channel inlet temperature (due to the operator's abrupt reduction of feedwater flow just over a minute earlier). The operators had thus taken the reactor from a situation where there was hardly any boiling in the core to a state of significant boiling. Because of the RBMK's positive void coefficient this meant an increase in reactivity and a rise in power. The fatal power excursion had begun.

Apparently in response to the observed increase in power, at 01.23.40 the shift manager ordered a full emergency shut down. The scram button, AZ-5, was pressed causing all rods to start motoring into the core. But it was too late.

Evidence that it was the power rise that prompted pressing of AZ-5 came from Legasov, who told the Vienna post accident meeting that "while operators were still alive, they told us that the moment when the scram button was pressed was when they noted power was increasing".

At 01.23.43 there were high power and short period alarms - the only trip channels left at this stage were those on short (10 seconds) period and on power overshoot (10 per cent). But, as manual shutdown had already been put into action, these were clearly of no help.

The reactor became prompt critical at 01.23.44, with power surging to well over 100 times full power. At about this time, severe shocks were felt and the operator noticed that rods were not inserting fully (possibly because of damage caused by power surge). Power to the control rod drive coupling mechanisms was cut to try to make the rods fall under gravity.

Fuel fragmented leading to a rapid increase in heat transfer to water and massive boiling. With cessation of flow into the core there was a second power surge to about 440 times full power at 01.23.45.

At 01.23.48 there was an explosion, due to build up of steam pressure, which blew the 1000t top shield off and rotated it through 90 degrees, rupturing all pressure tubes and destroying the reactor.

A few moments later a second explosion was heard, with a "firework display" of glowing particles and fragments emitted from the top of the reactor, causing about 30 fires. Several of these were on the roof of the turbine building, which is shared with the other three operating units at the site. The cause of this second explosion is believed by the Soviets to have been the combustion of hydrogen and carbon monoxide formed from the oxidation of zircaloy and $UO_2$ and from graphite/steam reactions.

Two people working in the reactor at the time of the accident died immediately. A number of other people died from radiation and burns within a fairly short time after the accident, many of them as a result of heroic efforts to get the fires under control. At the time of writing the Soviets put the death toll at 30. The figure has been revised down from the 31 that used to be reported on the grounds that one of the assumed victims, an old man whose body was found near the site, was found to have suffered a heart attack which according to the Soviets was not directly attributable to the accident.

## Why did it happen ?

The Chernobyl accident stemmed principally from a series of actions taken by those manning the plant. It attained catastrophic proportions because of design features of the RBMK reactor type. Underlying both the operational and design aspects of the accident, however, were deep seated organizational/institutional problems within the Soviet nuclear industry.

A remarkable feature of Chernobyl, which distinguishes it from, say, Three Mile Island, is that equipment failure played absolutely no part in the events that led to the accident.

Also noteworthy is that, among the actions taken by the operators, only one could be classified as human error in the usual sense — the mistake in setting control rods (at 00.28 on 26 April), which caused power to fall to a very low level. The other contributing actions were deliberate violations of operating rules in order to get the voltage regulator test completed no matter what. One reason for this single-mindedness may have been the fact that the window of opportunity for such a test (which, as already noted, should anyway have been done when the plant was initially commissioned) would only arise once a year when the plant was being taken down for its annual outage. There had also been the delay due to grid requirements. Furthermore it was the end of the working week and the middle of the night.

As it turned out the unintended fall in reactor power was crucial since it placed the reactor in the unfamiliar and problematic low power regime in which continuous operation is forbidden.

Similar voltage regulator tests had been done before in previous years at other plants, what distinguished this one were the difficulties created by being at low power. At low power the thermalhydraulics are unstable and difficult to control and, as noted already, the positive void coefficient dominates the negative fuel temperature (Doppler) coefficient — giving an overall positive power coefficient.

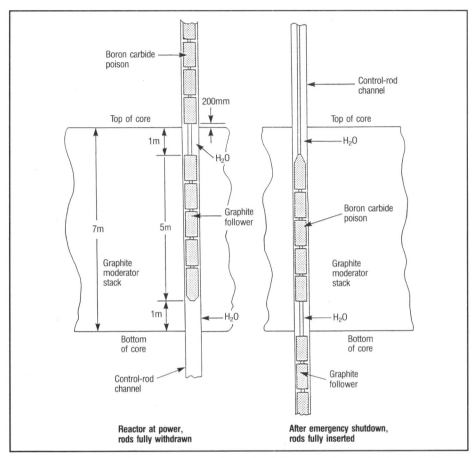

Boron carbide poison

Control-rod channel

200mm

Top of core

Top of core

1m

$H_2O$

$H_2O$

Graphite follower

Boron carbide poison

7m

5m

Graphite moderator stack

Graphite moderator stack

1m

$H_2O$

$H_2O$

Bottom of core

Bottom of core

Control-rod channel

Graphite follower

**Reactor at power, rods fully withdrawn**

**After emergency shutdown, rods fully inserted**

*When the control rods were inserted from the full out position, the initial effect would have been to increase reactivity in the lower part of the core due to displacement of water by the graphite followers attached below the boron carbide section of the control rods.*

In the attempts to raise power, control rods were withdrawn to the extent that the operating reactivity margin was grossly below the level the rules allowed. This was perhaps the most important of the several rule violations committed by the Chernobyl 4 operators. Having so few rods in the core, as well as increasing the time needed to shut down the reactor in emergencies (effectively rendering the reactor unprotected), also increases the positive void coefficient to a peak value.

Other major violations included blocking several reactor trip channels and having all eight main coolant pumps in operation at once, causing the entire cooling system to be on the verge of boiling.

The operators, as Legasov put it, "seemed to have lost all sense of danger". Complacency about safety, perhaps nurtured by Chernobyl's previous award winning performance record (it had previously been described as the flagship of the Soviet nuclear fleet), may have had something to do with it.

The fact that they were operating at significant power for nearly 12 hours with the ECCS disconnected (as a result of the delay due to grid requirements) is perhaps one reflection of the attitude that prevailed. Another illustration is that, not only was the test programme sloppily drawn up with a minimal regard for safety, but even then it was not properly carried through. There had been no agreement with the station physicists, the reactor builders. the RBMK designers, or representatives of the state nuclear safety inspectorate — although it had the approval of the station chief engineer.

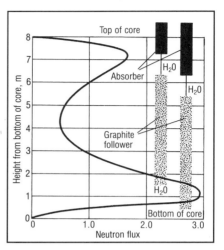

*Because of xenon poisoning and withdrawal of nearly all the rods, the axial flux shape was double humped, with the top and bottom of the core becoming to some extent decoupled.*

## Design deficiencies

While it is difficult to conceive of any current power reactor design which is not ultimately vulnerable to serious damage, given a sufficiently determined operating crew, the RBMK in the form being operated at Chernobyl 4 appears to have been particularly sensitive to operator malpractice and human failings in general.

Examples were the positive void coefficient and the possibility of the overall power coefficient becoming positive at low power*, the containment philosophy, the slow rate of emergency control rod insertion (taking up to about 18 seconds to reach the bottom of the core from the full out position), the absence of a fast shutdown system, and the apparent ease with which protective systems could be disabled.

---

* RBMK's have this undesirable characteristic primarily for reasons of improved fuel economy. To maximize the power obtainable from a given quantity of U-235, designers chose a ratio of graphite to water such that the graphite alone could moderate the nuclear reaction. In an RBMK the water's primary effect is that of absorber, so voiding will tend to lead to an increase in reactivity.

Overall, too much reliance was placed on the operators to keep their reactor out of trouble — "a colossal psychological mistake", according to Legasov. Quite a number of the operating rules were needed to counteract well-recognised design deficiencies, but very little had been done to help their enforcement through automated systems and operator support systems.

There is also the matter of "positive scram". It seems that when rods were inserted from the full out position (ie when AZ-5 was pushed), the initial effect would have been to increase reactivity in the lower part of the core due to displacement of water by the graphite followers attached below the boron carbide section of the control rods. As already noted, in the RBMK, water has an absorbing rather than a moderating effect, so where it is displaced by the followers, the effect is to increase reactivity. Normally this "positive scram" effect would not matter since the reactivity reduction towards the top of the core due to entry of the rods would overwhelm any reactivity increases elsewhere. However, at Chernobyl 4, because of xenon poisoning and withdrawal of nearly all the rods, the axial flux shape was double humped, with the top and bottom of the core becoming to some extent decoupled and acting as two independent critical masses.

The peak reactivity was at the bottom, amplifying the importance of the positive reactivity insertion due to displacement of water by the graphite followers.

While the existence of the positive scram characteristic in RBMKs of the Chernobyl 4 type is not in doubt (and presumably was one reason why the reactivity margin rule needed to be so strictly enforced), its precise contribution to the accident remains uncertain. Some argue that it was a key factor, others argue that it merely brought the inevitable runaway a few seconds forward, still others say it played no important role. A number of models of the accident do appear to suggest that to get an accident of Chernobyl proportions a large insertion of positive reactivity is needed at about the time that the AZ-5 button was pressed.

But the positive scram mechanism is only one of a number of possible sources of this extra reactivity. Another hypothesis is that cavitation caused sudden failure of the four main coolant pumps that were meant to remain in operation after the voltage regulator test (ie at around 35 seconds after the operators tripped the turbine, the voltage regulator cut out supplies to the four running down pumps, in accordance with the test plan, but due to a domino effect arising from cavitation the other four pumps ceased to function as well).

Another view is that the magnitude of the accident is fully explainable by void reactivity effects, with some amplification due to xenon burn out. The latter is essentially a positive feedback loop in which a rise in power causes more Xe-135 to be "burned out", ie converted to less absorbing isotopes, which in turn means an increase in power ... and so on.

The Soviets themselves have reported satisfactory modelling of the accident with their Triada 3-D code and are able to recreate reactor runaway so long as it assumed that around the time the AZ-5 button was pressed, there was what they describe as "electrical de-energization" of the main coolant pumps. This could refer to automatic cut-out of the pumps by the voltage regulator (that was under test) about 35 seconds after the operator tripped the turbine to start the experiment, or it could possibly refer to cut out of the other pumps (so that no pumps were left working), which would explain the large reactivity insertion.

## The institutional dimension

Looking behind these RBMK design deficiencies and the particular operator malpractices in the hours and minutes before the accident, a series of more fundamental institutional questions must inevitably be addressed. For example:

❑ Why were RBMK design deficiencies (apparently well known in the Soviet nuclear industry) not addressed more thoroughly and conscientiously ?

❑ How was it possible for control of a nuclear power plant effectively to pass to an electrical engineer who has been described as "not a specialist in reactors" ?

❑ Why was it possible for operating rules to be broken with such abandon, with the "deputy chief engineer" at the plant among those most responsible for the violations — to the extent that the impression is given that rule-breaking was a commonplace in the Soviet nuclear industry ?

❑ How was it possible for a test programme to be drawn up and to be implemented with, it would appear, complete bypassing of any sort of review and approvals process ?

❑ Where was the Soviet nuclear safety regulatory system in all of this ?

The accident brought to the surface serious nuclear safety management problems at the highest levels in the Soviet industry. The role of these institutional failings was recognised from the outset by the Soviets. One of the first steps they took after the accident was to set up a separate Ministry of Nuclear Energy to wrest nuclear power from what Prime Minister Nikolai Ryzhkov has called the "negative influence" of the Ministry of Power and Electrification, which at the time of the accident had responsibility for all power stations including nuclear ones.

The memoirs of Valery Legasov, published in Pravda shortly after his suicide on 27 April 1988 (the second anniversary of the accident), indicate very clearly the significance of institutional failures in the Chernobyl accident.

Legasov drew attention to what he felt was an over-complex and confusing organizational structure whereby there existed a collective rather than an individual responsibility for quality. This led to a "great deal of irresponsibility, as shown by the Chernobyl experience".

In reactor construction, Legasov cited a number of instances where production considerations overrode quality concerns. For example, there was the case of a shoddily welded pipe: "They began to look into the documentation and found all the right signatures: the signature of the welder, who certified that he had properly welded the seam, and the signature of the radiographer who had inspected the seam — the seam that had never existed. All this had been done in the name of labour productivity."

He also points out that the reactor protection system of the RBMK was recognised as being defective and proposals had been made as to how it might be improved but "not wishing to get involved in urgent additional work, the designer was in no hurry to change the protection control system".

Furthermore, he noted that the attitude to nuclear safety manifested by the experimenters at Chernobyl was not uncommon: "Experiments were carried out, the programme for which had been drawn up in an extremely negligent and untidy way, and there were no rehearsals of possible emergency situations in advance of experiments, disregard for the viewpoint of designers and nuclear physicists was total, and the correct fulfilment of procedures was something that had to be fought for".

In the case of Chernobyl itself, he referred to transcripts in his safe of conversations between operators which took place the night before the accident, which show that they were planning to carry out actions which had been crossed out in the operating manual. Legasov concludes despairingly: "The level of preparation of serious documents for a nuclear power plant was such that someone could cross out something, and the operator could interpret, correctly or incorrectly, what was crossed out and perform arbitrary operations".

He also observed that no attention was paid to the state of equipment until it was time for planned preventive maintenance and he recalled that one station manager had actually said: "What are you worried about? A nuclear reactor is only a samovar".

A generation of engineers had grown up in the Soviet nuclear industry who lacked any sort of critical attitude to the technology they were handling.

Alexander Kovalenko, now head of public information at Chernobyl, has confirmed (*Soviet Weekly*, 19 August 1989) that it had indeed become standard practice to break rules at Soviet nuclear plants. Station director Viktor Bryukhanov and his staff had developed the view that "instructions were for idiots"; understandable, as Kovalenko notes, in an environment where operation of plants was officially condoned before safety tests (such as that on the regulator itself) had been carried out. The operators also thought that "no matter what you did with the reactor an explosion was impossible", Kovalenko says. The operators were aware of a rule about having a minimum number of rods in the core, but did not know there could be an explosion. The possibility of explosions was not countenanced in the operating rules.

Kovalenko also notes that a sequence of events similar to the early stages of the Chernobyl accident had occurred at the Leningrad RBMK plant, and while the Chernobyl management had probably read the documents, "at the time of the tests they were all tucked up in bed".

The excessive importance which used to be attached to the production imperative at the expense of safety has been underlined by Mikhail Umanets, who is now director of the Chernobyl plant. He has recently made the revealing comment that: "Whereas we formerly regarded staff permitting the shutdown of a unit almost as people who had committed an offence, we now demand that a unit be shut down if there is the slightest doubt or lack of clarity".

The principal institutional features of the Chernobyl accident can therefore be summarized as follows:

- ❏ The organizational structure was such that safety responsibility was collective rather than individual ("safety management by committee").
- ❏ Existing hazards had been identified but were not acted upon.
- ❏ Concerns about "production" or "meeting the schedule" overrode safety and quality considerations.
- ❏ The safety envelope of the equipment was not well understood by those managing the operation of the plant.

## The wider issues

Looking to even wider underlying institutional issues, Legasov described the accident as the "apotheosis of all that was wrong in the management of the national economy and had been so for many decades", while Prime Minister Ryzhkov said shortly after the accident "we were *all* heading for a Chernobyl".

This notion that the accident was the outcome of institutional failings on a grand scale has prompted Alexander Kovalenko to argue that Viktor Bryukhanov, who was station manager at the time of the accident and is now serving a ten year prison sentence, has taken an unfair share of the blame.* Kovalenko contrasts the fate of Bryukhanov, who had succeeded in building the best nuclear plant town

---

* Viktor Bryukhanov was one of the three main defendants at the trial (which took place in July 1987), the others being Nikolai Fomin, chief engineer, and Anatoly Djatlov, deputy chief engineer (who was at the plant during the accident). All three received ten-year sentences. Other defendants were Alexander Kovalenko (head of reactor operations), who received a three-year sentence, Boris Rogoshkin (head of the night shift), five years, and Yuri Laushkin of the state nuclear safety inspectorate, two years. The trial heard that there had been an atmosphere of "lack of control and lack of responsibility", with staff playing cards and dominoes and writing letters. It also became clear that there had been several occasions in the past when accidents had been only narrowly avoided. Other sackings arising from the accident were Y. Kulov, chairman of the state nuclear safety inspectorate, G. A. Shasharin, deputy minister of power engineering and electrification, A. Meshkov, first deputy minister of medium machine building and V. S. Yemelyanov, deputy director of the Energotechnika Instute that designed the RBMK. A. Mayorets, minister of power and electrification received only a rebuke having been in post for only a short time.

in the country and who had taken a number of initiatives to improve training, with that of several senior party officials who at the time of the accident had responsibilities in the nuclear energy area and at the plant itself and who have since gone on to receive honours and decorations.

Kovalenko believes a key problem was that the party was taking decisions relating to nuclear energy but it was people like Bryukhanov who were having to answer for the consequences.

"The explosion, albeit at the best station in the country — was a natural outcome of events ... With our system of management and decision-making, a nuclear plant could not be safe. Therefore the main conclusion is: it was not the station that exploded, but our administrative, bureaucratic system. It is vital that everyone should answer for his actions. For that it is necessary to define precise limits to powers and duties. It must be made clear who answers for what."

Mikhail Gorbachev has also suggested that Chernobyl and the accidents that seem to have plagued the USSR in recent times are symptoms of a wider malaise in Soviet society. "We are haunted by accidents", he said, following the Urals gas pipeline disaster in June 1989. "Behind many of them there is negligence, irresponsibility, and lack of proper organization. It is impossible to disregard things of this kind. They keep happening too often."

Nevertheless, with glasnost and perestroika, the USSR is now endeavouring to build what may be a less accident-prone institutional environment.

One encouraging sign is a new found willingness to face up to past mistakes. For example, in a recent *Izvestia* interview Vladimir Asmolov of the Kurchatov Institute contrasted the reaction in the West to the Three Mile Island accident with that in the USSR. While Western countries massively increased spending on nuclear safety, the Soviet view was that "nuclear power is safe because it is safe", and no increased resources were devoted to safety. This was a product of the "years of public apathy and economic stagnation". Asmolov. said.

## Bibliography

P W W Chan, A R Dastur, S D Grant, J M Hopwood and B Chexal, *The Chernobyl accident: multidimensional simulations to identify the role of design and operational features of the RBMK 1000,* International Topical Conference on Probabilistic Safety Assessment and Risk Management, Zurich, September 1987.

P W W Chan, A R Dastur, S D Grant and J M Hopwood, *Chernobyl and pump cavitation (letter to the editor),* Nuclear Engineering International, August 1988.

J H Gittus, et al, *The Chernobyl accident and its consequences,* UKAEA, London, April 1988 (second edition).

J Q Howieson and V G Snell, *Chernobyl — a Canadian perspective, AECL 9334,* Atomic Energy of Canada Ltd, Candu Operations, Mississauga, Ontario, 1987.

INSAG (International Nuclear Safety Advisory Group), *Summary report on the post accident review meeting on the Chernobyl accident*, IAEA, Vienna, 1986

A Kovalenko, interview with V Medvedev, Soviet Weekly, 19 August 1989.

*Legasov suicide leaves unanswered questions*, Nuclear Engineering International, July 1988.

J C Luxat and B W Spencer, *Insights into the phenomenology and energetics of reactivity-initiated accidents*, Nucsafe 88, Avignon, 2–7 October 1988.

R F Mould, *Chernobyl: the real story*, Pergamon, Oxford, 1988.

L Nordstrom, *Was pump cavitation the key to Chernobyl?*, Nuclear Engineering International, May 1988.

*Soviet Chernobyl model highlights role of pump trip*, Nuclear Engineering International, May 1988

US Nuclear Regulatory Commission, et al, *Report on the accident at Chernobyl, NUREG 1250 Rev 1*, US NRC, Washington, DC, December 1987.

USSR State Committee on the Utilization of Atomic Energy, *The accident at the Chernobyl nuclear power plant and its consequences, information compiled for the IAEA post accident meeting,* 25–29 August 1986.

A White, *Consensus growing on positive scram at Chernobyl*, Nuclear Engineering International, January 1988.

# ASPECTS OF INSTITUTIONAL FAILURE

The accidents discussed here seem to have elements of institutional failure — that is, contributing causative factors which cannot be categorized in the classical fashion as "mechanical failure", "human error" or "catastrophic natural event".

These causative factors can be divided up into five interrelated categories of "failure", the failure in each case being a type of "human error": dominating production imperative; failure to allocate appropriate or adequate resources; failure to acknowledge or recognise an unsatisfactory or deteriorating safety situation; lack of appreciation of the technical safety envelope; and a failure to define and/or assign safety responsibility clearly.

The judgemental connotations of the words "error" and "failure" are often unfair in the context of the reactor accidents discussed here — this is especially the case for very early accidents such as NRX and Windscale where the contemporary state of the art was rudimentary and the whole nuclear energy enterprise was at an exploratory stage.

In the case of Three Mile Island, on the other hand, the "failures" identified could be more fairly attributed to the nuclear industry as a whole in the United States (including the regulatory authority), rather than just being laid on the shoulders of a single utility.

## Dominating production imperative

This term is used to describe not only the pressure to maintain production in a commercial/industrial environment, but also general schedule imperatives — the pressure to "get the job done on time" — or any sense of urgency that might affect a particular work programme or operation. The production imperative is clearly an essential discipline in any enterprise. Equally clearly it must not be allowed to override or dilute safety considerations to the extent that an unacceptable operating state results, nor can the question of balancing these two conflicting imperatives be completely delegated to the station manager — it must be a senior level responsibility in the operating institution.

The suggestion that production and safety conflict may seem at odds with the received wisdom, most notably articulated in INSAG-3, but as Weaver pointed out in his paper at the 1989 Quality in Nuclear Power Plant Operation Symposium in Toronto, the INSAG denial of a conflict is "brief, unsupported and not convincing", and indeed seems to contradict the evidence of a number of high consequence nuclear and non-nuclear events.

In addition, Swanson of the Electric Power Research Institute has pointed out a number of dangers inherent in the assumption that "safety" can be equated with "availability". In fairness to the people who framed INSAG-3 it is certainly unarguable that safety and long term economics must drive in the same direction — the costs of a major accident are too large for this to be otherwise — but this axiom should by no means be extended to become the statement that a reliable (ie high capacity factor) nuclear plant is necessarily a safe plant. Reliability of production, unless recorded over the total economic life of the plant (say 30 to 40 years), should not be regarded as the principal safety indicator.

Ballard identified the influence of the production imperative on the man-machine interface at an international conference on the man-machine interface in Tokyo in 1988. In the course of his analysis of errors by plant staff he pointed to a "conflict between the safety objective... and other objectives which may be economic". At the 1989 Toronto Symposium Frederick drew attention to the significant impact of the phenomenon on the quality of work performance (and, consequently, plant performance).

A dominating production imperative may be identified as a factor in the Windscale, SL-1 and Chernobyl accidents. The Windscale piles were, of course, part of an urgent and energetic programme for the development of nuclear weapons and it would be unrealistic to expect such energy and urgency not to be reflected in the approach to operations. Operating procedures, the Penney Report notes, evolved on a relatively informal basis and without adequate technical review, with the result that there was a general tendency to push pile temperatures progressively higher. The Report also mentions the very heavy work demands on the UKAEA's Industrial Group which effectively prevented adequate attention being paid to ensuring safe operation.

In the case of SL-1, Thompson points to a number of indications of the production imperative becoming dominant. He notes that "pressures of schedule and developmental difficulties" dictated the expedient of incorporating burnable poison in the form of discrete boron strips, spot-welded to the fuel elements, rather than dispersed in the fuel matrix. Examining the selected design, Thompson also points out that the vulnerability of a design which allowed criticality on withdrawal of the central rod was recognised, since that particular rod was provided with a longer follower than its fellows in order to prevent the rod falling completely out of the core should the rod drive fail. The fact that such recognition did not extend to evolving safer assembly/dissassembly means for the central rod may well be attributable to considerations of "extra cost or, more likely, time schedule". Though Thompson emphasises that this can only remain "conjecture", he does add: "The immediacy of a schedule delay or increased costs often outweighs the threat of a vague and improbable possibility".

The short time interval for transfer of operating responsibility of SL-1 from ANL to Combustion Engineering (three months) is another indicator of a "production imperative", Thompson noting that this does not appear to be an "adequate period".

In the case of Chernobyl a "production imperative" seems the only reasonable inference to draw from the unremitting determination of the operators, as documented in the USSR Report to the IAEA, to complete a planned test come what may. More generally, and informally, the late Academician Legasov's Memoirs quite explicitly identified the phenomenon in the context of the USSR nuclear power programme.

## Failure to allocate adequate or appropriate resources

In the examples considered here, "resources" principally refers to personnel, particularly in the technical support area, rather than material or equipment.

The NRX accident provided a demonstration of the need for the establishment of comprehensive written operational procedures and was a good illustration of the need for technical safety support for reactor operation. In fact, the formation of the "Safeguards Group" (subsequently to become the Reactor Technology Branch) at the Chalk River Nuclear Laboratories after the NRX accident not only met this need but also formalised responsibility for operational reactor safety.

The Penney Report identifies both material and personnel/technical resource deficiencies in the Windscale accident. Among the former, the most significant was the reactor instrumentation with respect to thermocouple positioning — under Wigner release conditions the in-core thermocouples would not be located at the highest temperature positions. However, a more serious omission was the almost complete lack of formal operating documentation for the reactor, the only guidance available being a 12 line memorandum, committee minutes and tradition.

Generally, technical support for Windscale operations was inadequate, a particularly important example of this being the failure to provide technical follow-up to previous Wigner release operations.

In the case of SL-1, resources were clearly inadequate in the areas of documentation and staffing. Rapid hand-over of the reactor to C-E and reassignment of design and start-up groups to other projects meant that C-E started its task with documentation that was almost certainly less than adequate — Thompson draws particular attention to this factor. Further, provision of round the clock professional supervision at SL-1 was proposed but rejected because of budgetary considerations.

The Fermi-1 accident provides a rather ironic example of failure to provide appropriate resources in that the operating institution actually provided a material "resource" at the suggestion of the ACRS (zirconium liner plates on the conical flow guide) but failed to provide resources to document this modification — a notable omission. In addition, no resources seem to have been applied to evaluating the desirability or otherwise of this modification.

At Lucens there was a failure to ensure continuity of supply of the crucial seals on the main coolant system blowers. In addition, a request by operations for a periscope to inspect the fuel channels was refused on economic grounds during the construction stage.

At Three Mile Island, inadequate resources, both material and human, can be identified as significant factors. The Kemeney Report draws specific attention to a lack of staff and expertise in the area of nuclear power plant operation, for example review of technical information from other plants was carried out by people without nuclear backgrounds. Other indicators of a paucity of human resources include the long-term lack of maintenance cited in the Report and deficiencies in Metropolitan Edison's implementation of its own quality assurance plan.

The failure to provide adequate operator training and the deficiencies in reactor instrumentation and control room design were also factors of immediate significance, although it is important to observe that from the changes made throughout the US nuclear industry since the Three Mile Island accident these "failures" were almost certainly not unique to the institution operating TMI. The Kemeny Report is careful to point out that Nuclear Regulatory Commission (NRC) standards allowed a low level of operator training and that there existed no NRC criteria for the qualifications of those conducting training programmes. Nevertheless, the Report makes it clear that the operator training programme for which Metropolitan Edison had primary responsibility was "qualitatively and quantitatively understaffed".

## Failure to acknowledge or recognize an unsatisfactory or deteriorating safety situation

Identification of "warning signals" is all too easy to do given the benefit of hindsight and the compilation of the appropriate documentation. However, the exercise is important because by identifying such missed warning signals methods can be evolved to increase the likelihood that future such signals may be detected, correctly interpreted and acted upon.

In the case of Windscale there was increasing difficulty in obtaining a successful release of Wigner energy. Of eight Wigner energy release operations carried out on Pile No 1 before the accident, two had achieved only partial releases and one had failed completely. Three of the later operations had required application of a second heating, initiated no less than 24 hours following the last graphite temperature rise. In recognition of the increasing difficulty of obtaining successful releases, the original 20 000 MWd interval between these operations had been extended to 40 000 MWd. But this recognition did not extend to establishing any formal written guidance, apart from a short memorandum, for those executing the operation, particularly with reference to the application of a second nuclear heating.

During the almost two-year period of operation of SL-1 a number of what USAEC General Manager C A Nelson called "undesirable conditions" developed in the reactor and, while unrelated to the immediate initiating event (the manual withdrawal of the central control rod), these conditions might well have prompted action to review SL-1 operations. Early concerns about the quality of documentation, particularly that relating to operation, did not, for example, result in any elaboration of the extremely laconic procedures for control-rod drive disassembly.

C-E's report to the AEC in 1959 (recommending the procurement of a new core) drew attention to the three vulnerable areas in the current reactor design, ie control rod drive design, control rod configuration and core materials. The recommendation was accepted by the AEC, yet despite implicit recognition that SL-1 was in a number of respects unsatisfactory no action seems to have been taken to reconsider operation. In August 1960 severe core deterioration was found, with clear safety implications, and was reported in the literature and an industry periodical. Despite this the reactor was returned to service at a higher power level than before without, apparently, any kind of comprehensive safety review. Towards the end of 1960 significant deterioration in control rod performance was experienced though this was not reported, even to the SL-1 project manager.

For Three Mile Island the most frequently cited "warning signal" was the 1977 Davis-Besse loss of feedwater transient. That this warning signal was not clearly transmitted represents a failure to respond to an unsatisfactory safety situation on the part of the regulatory authority. The Kemeney Report noted that in 1978 a Babcock and Wilcox engineer had identified the serious implications of this type of event and a Nuclear Regulatory Commission official had also pointed out the likelihood of inappropriate operator action. Similarly, Babcock and Wilcox not only failed to inform its customers of several incidents of stuck open PORVs but failed to highlight these failures in its own training programme.

The event which triggered the whole sequence of failures at Three Mile Island was the automatic isolation of the condensate polishers. Though repeated problems had been experienced with these polishers the Kemeney Report found that no effective steps had been taken to deal with them.

In addition, it should be noted that deficiencies in the alarm panels — which played an important role in the evolution of the accident — had been identified in an April 1978 memorandum by an operator who drew specific attention to the inadequate capacity of the alarm printer and the alarm system's over-sensitivity to feedwater transients.

## Lack of appreciation of the technical safety envelope

This particular failing occurring at senior institutional levels can underlie the three previous failings, in that a lack of understanding can result in:

❑ Demands for performance which overtax the capabilities of a particular technical system or the resources available to maintain it in safe operation.

❏ Failure to provide the necessary resources for the long-term safe
    operation of the system.
❏ Failure to understand the implications of operating experiences and to
    respond to them appropriately.

Lucens could certainly be considered to fall into this category. Quite apart from
the misapprehension concerning the implications of moisture in the primary
coolant system, new (and inadequate) blower seals were specified without
reference to operating experience.

In both the Three Mile Island and Chernobyl accidents there are indications of
this problem, albeit informal in the case of the latter. The Kemeney Report found
that, in the case of Three Mile Island, there existed a widespread lack of expertise
throughout the organisation. The Report noted that: "Nuclear power requires
management qualifications and attitudes of a very special character as well as an
extensive support system of scientists and engineers. We feel that insufficient
attention was paid to this..."

In a general and informal sense Academician Legasov has pointed out that in
the Soviet nuclear programme there certainly appeared a misapprehension about
the safety envelope at senior operational levels, citing a station manager's comparison of a reactor to a samovar. In the course of an appearance before the US Joint
Committee on Atomic Energy in June 1961 the late Admiral Hymann Rickover
(generally regarded as the architect of the US Navy's nuclear propulsion programme) addressed the problem of inadequate appreciation of the nature of
nuclear technology and agreed that it was difficult for many people to understand
that: "They are now dealing with a force that must be handled in an entirely
different manner than they have been accustomed". He also warned that the
nature of nuclear energy "is not thoroughly recognized by others who believe their
prerogatives to be invaded when someone suggests to them that if you do such
and such a thing you may incur danger. This idea is very difficult to get across."

## Failure to define and/or assign safety responsibility clearly

This seems to be a recurrent failure in the accidents described here. Clearly defined
safety responsibilities and authority are essential and must be reflected in the
organisational structure of any institution which employs technology with the
potential for high-consequence accidents. A necessary (but not sufficient) precon-
dition is that those at senior levels in the institution recognize their institution's
responsibility for safety (as articulated by Lord Marshall) and realize that ultimately
this responsibility must rest on their shoulders.

As was noted earlier, the NRX accident at Chalk River prompted the establishment of a technical group with assigned responsibility for safety. W B Lewis
noted in his report on the accident that although "design and management aimed

at setting conditions which would be safe despite normal human errors", the operating errors made at NRX were not "outside the normal range of human error" and "... a better system of review and inspection should be established. This should relate the design considerations to the current practice." To use the current phraseology, an important part of safety review should be to ensure that operating practice does not violate the design intent.

Responsibility for safety review must be clearly defined and structured so that this activity is distinct from "operation" or "production" responsibilities and derives its authority from an assigned senior authority in the institution. Lack of clear definition of safety responsibility and authority was clearly identified as a factor in the Windscale accident, the report drawing attention to the difficulties in ascertaining responsibility for particular technical decisions. In addition, it was pointed out that, because responsibilites were inadequately assigned and defined, there were not only some people in the organisation who might not be made aware of certain technical information necessary for the complete discharge of their duties, but also there were others who, while cognizant, might not appreciate that it was their responsibility to take such information into consideration in their work.

More generally, the Penny Report observed that "when the accident occurred the several responsibilities of the Chief Safety Officer, the Group Medical Officer and the Windscale Health Physics Officer were not clearly defined".

With the SL-1 operation the problem was a combined one of undefined responsibility and extreme organizational complexity. With the large numbers of organisations involved with varying degrees of responsibility for the project at different times there existed no single, well established channel up which all safety related information could flow for review by a single, competent and authoritative body. AEC staff inspections of the SL-1 project did not include operational safety and although several AEC Divisions had responsibility for or authority over various aspects of the SL-1 project none was formally assigned overall and continuous safety responsibility.

At his 1961 appearance before the USJCAE, Admiral Rickover was asked to comment on the problem of assignment of responsibility and the nature of responsibility for nuclear safety in the context of the SL-1 accident. Adopting the Navy propulsion programme as his example, the admiral noted: "The fact that a nuclear ship may go to sea for two months at a time, and be away from the United States, does not itself relieve me of my responsibility even though an operational commander has charge. I have been responsible for the design of the reactor plant, the training of the crew, the installation and test, the issuance of instructions. So no matter where an accident might happen, I am still personally responsible." He then noted: "Unless you can point your finger at the man who was responsible when something goes wrong then you have never had anyone really responsible".

As was noted in the chapter on the Fermi-1 fuel melting accident, lack of awareness of where safety responsibility ultimately falls may have been a factor in the events leading up to that event. The operating institution does not appear to have made significant efforts to determine whether or not installation of zirconium liner plates on the central flow guide would confer any palpable safety benefit, but simply followed a suggestion made by the Advisory Committee on Reactor Safeguards. An approach to safety design which can be colloquially summed up as "well, we've done what the regulatory authority wanted so we must be safe" represents a failure properly to understand safety responsibility, as that term is defined by Lord Marshall and articulated by Admiral Rickover.

The Kemeney Report pointed to indications of a similar problem in the relationship between operator and regulator when it identified "a preoccupation with regulations" noting that "the satisfaction of regulatory requirements is equated with safety". With specific reference to the Three Mile Island accident the Kemeney report noted other failures to recognise, define and assign safety responsibility adequately in both the regulatory authority and the operating utility. A "divided system of decision making" was noted within General Public Utilities and a number of examples were found where "this divided responsibility, in the case of TMI, may have led to less than optimal design and operating practices". The NRC was identified as having "serious managerial problems", "insufficient communication among the major offices" and "no well thought out, integrated system for the assurance of nuclear safety".

## Something more than a safety culture

The term "institutional failure" has been used to describe underlying causes of accidents or latent failures which seem to be traceable back to actions (or non-actions) at the senior management level which are informed by misunderstood, inadequate or non-existent safety related data (using that term in its broadest sense). The provisional definition cited in the Introduction, that is: "The impairment or absence of a corporate activity which is necessary for safety as a result of human error in activities which may not be acknowledged as important to safety and may occur far from the man-machine interface", is, as Weaver argues, too vague to be of much use in developing systematic means of detecting, correcting and minimizing the impact of the phenomenon. More study is needed to define more precisely exactly what it is that fails, how it fails and why it fails.

Organisational structure clearly plays an important role in institutional failure — for example, the excessively complex organisational structure of the SL-1 operation meant that establishment of a single, well defined and authoritative safety review process was a virtual impossibility. Pate-Cornell has drawn attention to the problems posed when a large number of organizational "layers" exist which can result in "degradation" of safety related information so that, for example, a "GO IF" signal can be stripped of its qualification to become a simple "GO" signal. Such degradation can be the result of safety related implications being unclearly expressed or even unidentified as such. Thus, not only does organizational structure have to be such that the number of "filters" are minimized, but also all corporate functions require examination to determine their relationship to, or impact upon, plant safety. Communication, then, becomes a related, vital factor. However, it is important to observe that in this context communication means more than unobstructed channels to senior levels in the institution. It includes comprehensive and unambiguous identification of safety-related issues and assurance that these are appropriately factored into an institution's policies and strategies.

At the 1989 Canadian Nuclear Society Annual Meeting, Weaver described a preliminary survey of the extent to which technical considerations inform senior level corporate decision making and reported that he had found no indication in the literature that technical input had a part to play in the formulation of strategic or policy decisions. Should further, more rigorous, examination confirm this then the implications may be disturbing.

The organisational structure surrounding nuclear power plant operation should reflect the plant's status as the utility's revenue earner and principal capital investment. It should also reflect the fact that a nuclear power plant is not an autonomous entity as far as safety is concerned, any more than, for example, is a railway train or a commercial airliner. This is by no means at odds with Thompson's observation that ultimate responsibility for nuclear reactor safety resides in the operating personnel at the reactor who must have the authority to shut it down should they believe it unsafe, but rather recognises the fact that responsibility and authority for safety in operation (as distinct from shut-down) extend beyond the plant boundary.

In INSAG-3 the International Nuclear Safety Advisory Group developed the concept of a "safety culture" which refers to "a very general matter, the personal dedication and accountability" of all those involved in any activities influencing nuclear plant safety. The report also identifies senior management as the "starting point" for the necessary full attention to safety matters.

The concept of a "safety culture" may be a welcome first step towards countering institutional failure, particularly with its emphasis on senior level responsibility for safety and the need to set in place a generalized "safety imperative". However, repeated invocation of the term "safety culture" (or, indeed, "institutional failure") will not of itself lead to much. What is needed is a serious attempt to examine exactly how institutions fail to discharge their safety responsibilities, with particular emphasis on both the inputs to management decisions with safety implications and the relative influences of those inputs.

As the range of papers at the Safety and Reliability Symposium held at Altrincham in 1988 shows, increasing attention is being focussed on this area with the study of "human factors" (in the widest sense of the term) now being extended beyond the man-machine interface.

As Weaver has suggested, another important development will be the evolution of quality standards to cover a range of corporate functions identified as having an influence on plant safety — a move in this direction, he notes, is being made in Canada by the Advisory Committee on Nuclear Safety, which has formulated a proposal for institutional quality assurance.

To counter institutional failure, therefore, a good "safety culture" is necessary but it is not by itself sufficient. Clear definitions of the safety elements of all corporate functions need to be formulated on an institution-specific basis.

Providing and maintaining protection against high-consequence, but infrequent, accidents is a cost and cost control is one of the responsibilities of senior management — as is safety itself. It seems unreasonable to expect senior management to discharge both safety and cost control responsibilities without any clear understanding of the impact of the latter on the former.

## Bibliography

*Basic Safety Principles for Nuclear Power Plants,* IAEA Safety Series No 75-INSAG-3, IAEA, Vienna, 1988.

*Proceedings of the International Symposium on Quality in Nuclear Power Plant Operation,* ISBN 0-919307-38-8, Toronto, 1989:

 — K R Weaver, *A Conceptual Study of the Role of the Institution in Safety.*
 — E R Frederick, *Breaking Down Quality Performance Barriers.*

R Swanson, *Safety is Not Equal to Availability,* Nuclear Safety, Vol. 23 No. 1, January/February 1982.

G M Ballard, *Reactor Events Involving Misinterpretation / Misunderstanding of Plant Status by Plant Staff,* Man-Machine Interface in the Nuclear Industry, STI/PUB/781, IAEA, Vienna, 1988.

Hearings Before the Joint Committee on Atomic Energy, Congress of the United States, Eighty-seventh Congress. First Session on Radiation Safety and Regulation, 13, 14, 15 June 1961, US Government Printing Office, Washington, 1961.

K Weaver, *A Review of Safety Considerations in Organizational Decision Making,* Annual Meeting of the Canadian Nuclear Society, Ottawa, June 1989.

T J Thompson and J G Beckerley, *The Technology of Nuclear Reactor Safety,* MIT Press, 1964. Chapter 11 "Accidents and Destructive Tests", (Thompson).

E Pate-Cornell, *Organizational Factors in Reliability Models,* Proceedings of the 1988 Meeting of the Society for Risk Analysis, Washington DC, November 1988